Other Titles in This Series

(Continued in the back of this publication)

MEMOIRS
of the
American Mathematical Society

Number 579

Lebesgue Theory
in the Bidual of C(X)

Samuel Kaplan

May 1996 • Volume 121 • Number 579 (third of 4 numbers) • ISSN 0065-9266

American Mathematical Society
Providence, Rhode Island

1991 *Mathematics Subject Classification.*
Primary 46G99, 28C05; Secondary 46B42.

Library of Congress Cataloging-in-Publication Data

Kaplan, Samuel, 1916 Sept. 13–
  Lebesgue theory in the bidual of C(X) / Samuel Kaplan.
    p. cm. – (Memoirs of the American Mathematical Society, ISSN 0065-9266; no. 579)
  "Volume 121, number 579 (third of 4 numbers)."
  Includes bibliographical references and indexes.
  ISBN 0-8218-0463-4 (alk. paper)
  1. Lebesgue integral. 2. Radon measures. 3. Banach lattices. I. Title. II. Series.
QA3.A57 no. 579
[QA312]
510 s–dc20
[515′.43]   96-2232
                                                                                CIP

Memoirs of the American Mathematical Society

This journal is devoted entirely to research in pure and applied mathematics.

Subscription information. The 1996 subscription begins with Number 568 and consists of six mailings, each containing one or more numbers. Subscription prices for 1996 are $391 list, $313 institutional member. A late charge of 10% of the subscription price will be imposed on orders received from nonmembers after January 1 of the subscription year. Subscribers outside the United States and India must pay a postage surcharge of $25; subscribers in India must pay a postage surcharge of $43. Expedited delivery to destinations in North America $30; elsewhere $92. Each number may be ordered separately; *please specify number* when ordering an individual number. For prices and titles of recently released numbers, see the New Publications sections of the *Notices of the American Mathematical Society.*

Back number information. For back issues see the *AMS Catalog of Publications.*

Subscriptions and orders should be addressed to the American Mathematical Society, P. O. Box 5904, Boston, MA 02206-5904. *All orders must be accompanied by payment.* Other correspondence should be addressed to Box 6248, Providence, RI 02940-6248.

Memoirs of the American Mathematical Society is published bimonthly (each volume consisting usually of more than one number) by the American Mathematical Society at 201 Charles Street, Providence, RI 02904-2213. Second-class postage paid at Providence, Rhode Island. Postmaster: Send address changes to Memoirs, American Mathematical Society, P. O. Box 6248, Providence, RI 02940-6248.

CONTENTS

CONTENTS

ABSTRACT

The present work is based upon our monograph The Bidual of C(X) I (X being compact). We generalize to the bidual the theory of Lebesgue integration, with respect to Radon measures on X, of bounded functions. The bidual of C(X) contains this space of bounded functions, but is much more "spacious", so the body of results can be expected to be richer. Finally, we show that by projection onto the space of bounded functions, the standard theory is obtained.

Key words. Banach lattice, Radon measure, unbounded order convergence, almost everywhere.

Received by the editor October 27, 1994.

Introduction $-1 - 7 \ (\S 1)$

In this paper, X is a fixed compact space, C is the space of real continuous functions on X, C' its dual, and C'' its bidual, with $C \subset C''$. C, C', and C'' are Banach lattices, and our treatment is based on this characterization of them.

The elements of C' are the Radon measures on X. As is well known, they can be identified with the regular measures on X. For each $\mu \in C'$, $\mu \geq 0$, the band of C' generated by μ can be identified with $L^1(\mu)$. Thus the $L^1(\mu)$'s, μ running through C'_+, appear in a natural way as direct summands of C'. It follows their duals, the spaces $\{L^\infty(\mu) : \mu \in C'_+\}$ appear as direct summands of C''. The space $\ell^1(X)$ also appears as a direct summand of C', hence its dual $\ell^\infty(X)$ (the space of bounded functions on X) appears as a direct summand of C''. Under its canonical imbedding in C'', C is *not* contained in $\ell^\infty(X)$. It has been "lifted" to a central position in C'', and what seems to be C in $\ell^\infty(X)$ is actually its image under the projection of C'' onto $\ell^\infty(X)$. ‘The Bidual of C(X) I’ (NH 1985) — but it failed to recover.

In [3], we treated C'' as a generalization of $\ell^\infty(X)$, and starting with C, obtained the Baire elements of C'', the lsc and usc elements ("lsc" and "usc" to suggest lowersemicontinuous and uppersemicontinuous), the Borel elements, etc.. And for each $\mu \in C'_+$, we obtained the elements of C'' "Riemann integrable" with respect to μ. The projection of each of these families on $\ell^\infty(X)$ gives the standard bounded Baire functions, bounded lowersemicontinuous functions, etc. on X. However, since C'' is much more "spacious" than $\ell^\infty(X)$, the imbedding of our families in C'' is different than that of the corresponding families in $\ell^\infty(X)$, and the body of results in C'' is richer than that in $\ell^\infty(X)$. An example: In standard analysis, with C a subspace of $\ell^\infty(X)$, every $f \in \ell^\infty(X)$ is the limit under order convergence of some net in C. In C'', the set of limits under order convergence of nets in C is a proper subspace U of C''. Its image, under projection into $\ell^\infty(X)$ is precisely the set of functions on X Lebesgue integrable with respect to every $\mu \in C'_+$ (hence the notation U for "universally integrable"). U turns out to be the most important subspace of C'' after C.

In the present paper, we study Lebesgue theory in C'', that is, for each $\mu \in C'_+$, we define for elements of C'', "Lebesgue integrability" with respect to μ and study its properties. And again, the projection of the set of "Lebesgue integrable" elements into $\ell^\infty(X)$ is precisely the set of bounded μ-integrable functions.

In C'', the clasical theorems of integration theory — for example, the Lebesgue Bounded Convergence Theorem, the Egorov theorem, and the Lusin Theorem — are obtained quite simply from the Banach lattice properties of C'', with the bilinear form $\langle f, \mu \rangle$ on

$C'' \times C'$ playing the role played by $\int f d\mu$ in $\ell^\infty(X)$.

We note an additional example of the greater spaciousness of C''. For each $f \in \ell^\infty(X)$, there exist two elements f_*, f^* in C'', both projecting onto f, such that for every $\mu \in C'_+$ — denoting the upper and lower Darboux integrals by $\int^* f d\mu$ and $\int_* f d\mu$ — $\langle f^*, \mu \rangle = \int^* f d\mu$ and $\langle f_*, \mu \rangle = \int_* f d\mu$.

§1. Preliminaries

We recall definitions and results from [3]. For terminology and notation not specifically defined, we refer the reader to [3] or any book on Riesz spaces. $\mathbb{N}$ denotes the natural numbers, $\mathbb{R}$ the reals. n, m will always denote elements of $\mathbb{N}$, and λ, κ, ρ elements of $\mathbb{R}$. **1** denotes the element of C with constant value 1 on X. It is the (Banach lattice) *unit* for C, that is:

(1) it is a strong unit: for every $f \leq C$, $|f| \leq \lambda \mathbf{1}$ for some λ;

(2) for each $f \in C''$, $\|f\| = \inf\{\lambda: \lambda$ satisfies (1)$\}$.

Subsets of C'' will be denoted by upper case letters or by $\{f_\alpha\}$, it being understood that the index runs through some set $\mathcal{A}$. However, when we speak of the *net* $\{f_\alpha\}$, this will mean that $\mathcal{A}$ is a directed set, that is, it has an order such that for every $\alpha_1, \alpha_2 \in \mathcal{A}$, there exists $\alpha_3 \geq \alpha_1, \alpha_2$. $f_\alpha \longrightarrow f$ will mean that the net $\{f_\alpha\}$ order converges to f (so contains implicitly that $\{f_\alpha\}$ is order bounded).

A subset of C'' is *order closed* (resp. *σ-order closed*) if it is closed under order convergence of nets (resp. sequences). The *order*

✗ O is unnatural — see p. 28.

closure (resp. *σ-order closure*) of $A \subset C''$ is the smallest order closed (resp. σ-order closed) set containing A. A linear functional ϕ on C'' is *order continuous* on C'' if for every net $\{f_\alpha\}$ in C'', $f_\alpha \longrightarrow f$ implies $\langle f, \phi \rangle = \lim_\alpha \langle f_\alpha, \phi \rangle$. And similarly for a linear functional on C', C'' is the space of all order continuous linear functionals on C', and dually, C' is the space of all order continuous linear functionals on C'' (in Köthe theory, C' and C'' are *perfect*).

A *band* of C' (resp. C'') is an order closed Riesz ideal. C' and C'' are Dedekind complete, so in both, bands are direct summands. In detail: for each band J of C', the *disjoint* of J, $J^d \doteq \{\nu \in C':$ $|\nu| \wedge |\mu| = 0$ for every $\mu \in J\}$ is also a band, and $C' = J \oplus J^d$; and similarly for C''. Given a band G of C'', it follows from $C'' = G \oplus G^d$ that every $f \in C''$ has a unique component in G; we denote it by f_G. We call the mapping $f \mapsto f_G$ ($f \in C''$) the *projection* of C'' onto G. And for each $A \subset C''$, we denote by A_G the image of A under the projection of C'' onto G: $A_G = \{f_G: f \in A\}$. The same discussion holds for a band of C'.

For each band J of C', we denote by $J^\perp$ its null space in C''; and for each band G of C'', we denote by $G_\perp$ its null space in C'. Given a band J of C', $J^\perp$ is itself a band (of C''), so $C'' = (J^\perp)^d \oplus J^\perp$. We will call $(J^\perp)^d$ the band of C'' *dual* to J. Denote it by I. (Note that $I = (J^d)^\perp$ also.) The relation is symmetric: $J = (I_\perp)^d$. And if we start with a band I of C'' and set $J = (I_\perp)^d$, then $I = (J^\perp)^d$. The notation (I, J) will always denote a pair related in this way: $I = (J^{\perp})^d$ and $J = (I_{\perp})^d$, and we will call such a pair of bands a *dual pair* of bands.

A dual pair (I, J) inherits the duality relations of the pair (C'', C'). I is the norm dual of J and also the space of order continuous linear functionals on J; and J is the space of order continuous linear functionals on I. (It is not the norm dual, only the predual.)

We have mentioned an important dual pair in the introduction $(\ell^{\infty}(X), \ell^1(X))$. In detail: We consider X as imbedded in C'. Each $x \in X$ is the element of C' defined by $\langle f, x \rangle \doteq f(x)$ for every $f \in C$. The norm closed linear subspace of C' generated by the subset X can be identified with $\ell^1(X)$. It turns out that this linear subspace is actually a band of C'. Its elements are the *atomic* (Radon) measures on X; to indicate this, we denote it by C_a'. The disjoint $(C_a')^d$ consists of the *diffuse* (Radon) measures, and we denote it by C_d'. Thus $C' = C_a' \oplus C_d'$. we denote the bands of C'' dual to C_a' and C_d' by C_a'' and C_d'' respectively. So $C'' = C_a'' \oplus C_d''$. For each $f \in C''$, we will denote the components of f in C_a'' and C_d'' by f_a and f_d (instead of $f_{C_a''}$ and $f_{C_d''}$). And for $A \subset C''$, we will write A_a for $A_{C_a''}$ and A_d for $A_{C_d''}$.

Since C_a'' is the norm dual of $C_a' = \ell^1(X)$, it can be identified with $\ell^{\infty}(X)$. We remind the reader again that $C \not\subset C_a''$, that what looks like C in C_a'' is actually C_a.

An important family of dual pairs of bands consists of those of the form $(L^\infty(\mu),\ L^1(\mu))$. Consider $\mu \in (C')_+$. We denote by C'_μ the band J generated by μ, and by C''_μ the band I of C'' dual to it. As is well known (via the Radon-Nikodym theorem), C'_μ can be identified with $L^1(\mu)$, so its dual, C''_μ, can be identified with $L^\infty(\mu)$. For each $f \in C''$, we will denote its component in C''_μ by f_μ (instead of $f_{C''_\mu}$) and for each $A \subset C''$, we will denote $A_{C''_\mu}$ simply by A_μ.

We recall some additional definitions from [3]. Given $A \subset C''$, A^ℓ is the set of suprema in C'' of subsets of A, and A^u the set of infima. (If we take $A = C$, the elements of C^ℓ are the lsc elements — their images in C''_a under the projection of C'' onto C''_a are the (bounded) lower semicontinuous functions on X. Similarly, the elements of C^u are the usc elements. This is why we use the notation A^ℓ and A^u.) The *Dedekind closure* of A in C'' is $A^\ell \cap A^u$, the set of elements which are each the supremum of some subset of A and the infimum of some subset of A. A is contained in its Dedekind closure; if they coincide, we will say A is *Dedekind closed* in C''.

The *Baire* subspace $\mathfrak{Ba}$ of C'' is the σ-order closure of C in C''. The *Borel* subspace $\mathfrak{Bo}$ is the σ-order closed linear subspace generated by C^ℓ (equivalently, by C^u). We have

$$C \subset \begin{Bmatrix} \mathfrak{Ba} \\ C^\ell \\ C^u \end{Bmatrix} \subset \mathfrak{Bo} \subset U,$$

where U is the subspace introduced in the Introduction as the set of limits under order convergence of nets in C.

U will play a central role in the present paper. We list those of its properties which we will be using.

(U1) U is a Riesz subspace of C'' containing C.

(U2) U is σ-order closed in C''.

(U3) U is Dedekind closed in C''.

(U4)   the projection of U onto U_a is a (Banach lattice) isomorphism.

(U5) $U \cap C''_d = 0$ (this follows from (U4)).

(U6) U consists of those elements of C'' which are each the infimum of a set of lsc elements and the supremum of a set of usc elements.

CHAPTER 1. $\mathfrak{L}^{\infty}$

§2. The upper and lower "envelopes" of an element

In [3], starting from the subspace C, we defined the Riemann integrable elements of C'' and studied their properties. In the present paper, we follow the the same procedure to obtain the Lebesgue integrable elements and their properties, but this time

U – p.2.

starting from U instead of C. The work could be done starting from the space Bo of Borel elements, but we believe U is more natural.

The important difference, making Lebesgue theory superior to Riemann theory is that U is σ-order closed in C'', while C is, in general, not.

Our first step in [3] was to define for each $f \in C''$, the "upper envelope" $u(f)$ of f, and the "lower envelope" $\ell(f)$, by

$$u(f) = \wedge \{h \in C \colon h \geq f\},$$
$$\ell(f) = \vee \{g \in C \colon g \leq f\}.$$

Thus $u(f)$ is the smallest usc element above f and $\ell(f)$ is the largest lsc element below it.

Remark. Note the similarity with topology. The lsc elements correspond to the open sets, the usc elements to the closed ones, $u(f)$ to the closure of a set, and $\ell(f)$ to the interior.

We now parallel the above. First, some properties of U^{ℓ} and U^u. A subset of a vector space is called a *wedge* if it is closed under

8

addition and multiplication by elements of $\mathbb{R}_+$.

(2.1)   (1) U^ℓ is a wedge closed under the operations of taking suprema of arbitrary subsets and infima of countable ones.

(2) U^u is a wedge closed under the operations of taking infima of arbitrary sets and suprema of countable ones.

It follows U^ℓ and U^u are σ-order closed.

For a proof, cf. [3; (42.1)].

Remarks. (1) The closedness of U^ℓ under the taking of infima of countable subsets is not true for C^ℓ, and the closedness of U^u under the taking of suprema of countable subsets is not true for C^u. Thus the final statement does not hold for C^ℓ and C^u.

(2) U^ℓ can also be defined as the set of suprema of sets of usc elements, and U^u can be defined as the set of infima of sets of lsc elements (cf (U6) at the end of §1).

It is to be expected that U^ℓ and U^u are larger than C^ℓ and C^u.

(2.2)   (1) $(C''_a)_+ \subset U^\ell$.

(2) $U^\ell \cap (C''_d)_+ = 0$.

(3) For each $\mu \in C'_+$, $(C''_\mu)_+ \subset U^u$.

Proof. (1)   Consider $f \in (C''_a)_+$. For each $x \in X$, the element of C''_a which coincides with f at x and vanishes elsewhere is a usc elements [3; (54.13)], hence lies in U. Since f is the supremum of the set of all such elements, f lies in U^ℓ.

(2)   This follows from the fact that $U \cap C_d'' = 0$ (Property (U4)).

(3) This is the Lemma in the proof of [3; (43.2)].

$$\text{QED}$$

We will call the elements of U^ℓ *$\boldsymbol{\ell}$-elements* and the elements of U^u *$\mathbf{u}$-elements*. (Note that $\boldsymbol{\ell}$ and $\mathbf{u}$ are boldface.)

Note that $U^u = -U^\ell$. A useful consequence is that if f is a u-element and g an $\boldsymbol{\ell}$-element, then $f - g$ is a u-element and $g - f$ is an $\boldsymbol{\ell}$-element.

For each $f \in C''$, we set

$$\mathbf{u}(f) = \wedge \{h \in U: h \geq f\},$$

$$\boldsymbol{\ell}(f) = \wedge \{g \in U: g \leq f\}.$$

Thus $\mathbf{u}(f)$ is the smallest u-element above f, and $\boldsymbol{\ell}(f)$ is the largest $\boldsymbol{\ell}$-element below f. All the properties of the operations $\mathbf{u}(\cdot)$ and $\boldsymbol{\ell}(\cdot)$ recorded in [3] hold for $\mathbf{u}(\cdot)$ and $\boldsymbol{\ell}(\cdot)$, but the latter have important stronger properties. We will omit proofs for $\mathbf{u}(\cdot)$ and $\boldsymbol{\ell}(\cdot)$ which are identical with ones given in [3] for $\mathbf{u}(\cdot)$ and $\boldsymbol{\ell}(\cdot)$.

(2.3)   (1) $\boldsymbol{\ell}(f) \leq \boldsymbol{\ell}(f) \leq f \leq \mathbf{u}(f) \leq \mathbf{u}(f)$ for every $f \in C''$.

(2) $g \leq f$ implies $\boldsymbol{\ell}(g) \leq \boldsymbol{\ell}(f)$ and $\mathbf{u}(g) \leq \mathbf{u}(f)$.

(3) For every $\rho \in \mathbb{R}_+$, $\boldsymbol{\ell}(\rho f) = \rho\boldsymbol{\ell}(f)$ and $\mathbf{u}(\rho f) = \rho\mathbf{u}(f)$.

(4) $\mathbf{u}(f) = \wedge \{h: h \text{ an lsc element } \geq f\}$,

$$\ell(f) = \vee \{g: g \text{ a usc element } \leq f\}.$$

(5) $u(-f) = -\ell(f)$ for every $f \in C''$.

The verifications are straightforward.

(2.4) For every $f, g \in C''$,

$$\ell(f) + \ell(g) \leq \ell(f+g) \leq \ell(f) + u(g) \leq u(f+g) \leq u(f) + u(g).$$

It follows

$$\ell(f) - u(g) \leq \ell(f-g) \leq \begin{matrix} u(f) - u(g) \\ \\ \ell(f) - \ell(g) \end{matrix} \leq u(f-g) \leq u(f) - \ell(g).$$

cf. the proof of [3; (49.1)].

If a set $\{f_\alpha\}$ in C'' is order bounded above, then so is the set $\{u(f_\alpha)\}$; and if it is order bounded below, then so is $\{\ell(f_\alpha)\}$. Note also than in C'', order boundedness and norm boundedness are equivalent, so we can use the term "boundedness" without ambiguity.

(2.5) For a bounded set $\{f_\alpha\}$ in C'':

(1) $u(\wedge_\alpha f_\alpha) \leq \wedge_\alpha u(f_\alpha) \leq \vee_\alpha u(f_\alpha) \leq u(\vee_\alpha f_\alpha)$.

(2) $\ell(\wedge_\alpha f_\alpha) \leq \wedge_\alpha \ell(f6) \leq \vee_\alpha \ell(f_\alpha) \leq \ell(\vee_\alpha f_\alpha)$.

cf. [3; (49.3)].

If $\{f_\alpha)$ is countable, then (unlike the case for $u(\cdot)$ and $\ell(\cdot)$) the last inequality in (1) and the first in (2) become equalities:

(2.6)   For a countable bounded set $\{f_n\}$ in C'':

   (1) $\vee_n u(f_n) = u(\vee_n f_n)$,

   (2) $\wedge_n \ell(f_n) = \ell(\wedge_n f_n)$.

In particular, for $f, g \in C''$:

$$u(f) \vee u(g) = u(f \vee g),$$
$$\ell(f) \wedge \ell(g) = \ell(f \wedge g).$$

Proof. We prove (1). Set $h = \vee_n u(f_n)$; we need only show $h \geq u(\vee_n f_n)$. $u(f_n) \geq f_n$ for every n, so $h \geq \vee_n f_n$. But by (2.1), h is a **u**-element, hence $h \geq u(\vee_n f_n)$.

QED

(2.7)   Given a bounded sequence $\{f_n\}$ in C'':

   (1) $\liminf_n \ell(f_n) \leq \ell(\liminf_n (f_n))$
   $$\leq u(\liminf_n (f_n))$$
   $$\leq \liminf_n u(f_n).$$

   (2) $\limsup_n \ell(f_n) \leq \ell(\limsup_n (f_n))$
   $$\leq u(\limsup_n (f_n))$$
   $$\leq \limsup_n u(f_n).$$

It follows that if $f_n \to f$, then

$$\limsup_n \ell(f_n) \leq \ell(f) \leq u(f) \leq \liminf_n u(f_n).$$

Proof. Set $g = \liminf_n (f_n)$ and $h = \limsup_n (f_n)$. Then the above can be written:

(1) $\liminf_n \ell(f_n) \leq \ell(g) \leq u(g) \leq \liminf_n u(f_n).$

(2) $\limsup_n \ell(f_n) \leq \ell(h) \leq u(h) \leq \limsup_n u(f_n).$

We prove (1). For each n, set $g_n = \wedge_{m \geq n} f_m$. Then

$$\liminf_n \ell(f_n) = \vee_n \wedge_{m \geq n} \ell(f_m)$$

$$= \vee_n \ell(g_n) \tag{2.6}$$

$$\leq \ell(\vee_n g_n) \tag{2.5}$$

$$= \ell(g);$$

and $$u(g) = u(\vee_n g_n)$$

$$= \vee_n u(g_n) \tag{2.6}$$

$$\leq \vee_n \wedge_{m \geq n} u(f_m)$$

$$= \liminf_n u(f_n).$$

QED

Of the equalities and inequalities obtained so far, the binomial ones can be sharpened when one of the elements lies in U.

(2.8)   If $g \in U$, then for every $f \in C''$:

$$\ell(f+g) = \ell(f) + g,$$

$$u(f+g) = u(f) + g.$$

This follows easily from (2.4) and the identity $\ell(g) = u(g) = g.$

(2.9) If $g \in U$, then for every $f \in C''$:

$$\ell(f \wedge g) = \ell(f) \wedge g,$$

$$u(f \wedge g) = u(f) \wedge g,$$

$$\ell(f \vee g) = \ell(f) \vee g,$$

$$u(f \vee g) = u(f) \vee g.$$

cf [3; (49.9)].

Setting $g = 0$, we obtain

(2.10) For every $f \in C''$:

$$(u(f))^+ = u(f^+),$$

$$(u(f))^- = \ell(f^-),$$

$$(\ell(f))^+ = \ell(f^+),$$

$$(\ell(f))^- = u(f^-).$$

To establish the inequality (2) below, we use the inequality $\ell(f \vee g) \leq \ell(f) \vee u(g)$. For a proof, cf. [3; (49.5)].

(2.11) For every $f \in C''$:

(1) $u(|f|) = |u(f)| \vee |\ell(f)|$,

(2) $\ell(|f|) \leq |u(f)| \wedge |\ell(f)|$.

Proof.

(1) $|u(f)| \vee |\ell(f)| = [u(f) \vee (-u(f))] \vee [\ell(f) \vee (-\ell(f))]$

$$= u(f) \vee \ell(-f) \vee \ell(f) \vee u(-f) \qquad (2.3)$$

$$= \mathbf{u}(f) \vee \mathbf{u}(-f) \tag{2.3}$$

$$= \mathbf{u}(f \vee (-f)) \tag{2.6}$$

$$= \mathbf{u}(\,|\,f\,|\,).$$

(2) (using (2.3) and the equality noted above)

$$\boldsymbol{\ell}(f) \vee \mathbf{u}(-f) = \boldsymbol{\ell}(f) \vee (-\boldsymbol{\ell}(f)).$$

$$\boldsymbol{\ell}(\,|\,f\,|\,) = \boldsymbol{\ell}(f \vee (-f)) \; \leq$$

$$\mathbf{u}(f) \vee \boldsymbol{\ell}(-f) = \mathbf{u}(f) \vee (-\mathbf{u}(f)).$$

Thus $\boldsymbol{\ell}(\,|\,f\,|\,) \; \leq \; |\,\boldsymbol{\ell}(f \vee \,|, \,|\,\mathbf{u}(f)\}.$

Hence $\boldsymbol{\ell}((\,|\,f\,|\,) \; \leq \; |\,\boldsymbol{\ell}(f)\,| \; \wedge \; |\,\mathbf{u}(f)\,|\,.$

QED

Contained in the above, we have

$$|\,\mathbf{u}(f)\,|$$

$$\boldsymbol{\ell}(\,|\,f\,|\,) \; \leq \qquad\qquad \leq \; \mathbf{u}(\,|\,f\,|\,).$$

$$|\,\boldsymbol{\ell}(f)\,|$$

(2.12) For $f, g \in C''$,

$$|\,\mathbf{u}(f) - \mathbf{u}(g)\,|$$

$$\leq \; \mathbf{u}(\,|\,f - g\,|\,).$$

$$|\,\boldsymbol{\ell}(f) - \boldsymbol{\ell}(g)\,|$$

Proof. $|\,\mathbf{u}(f) - \mathbf{u}(g)\,| \; = [\mathbf{u}(f) - \mathbf{u}(g)] \vee [\mathbf{u}(g) - \mathbf{u}(f)]$

$$\leq \; \mathbf{u}(f - g) \vee \mathbf{u}(g - f) \tag{2.4}$$

$$= \mathbf{u}[(f - g) \vee (g - f)] \tag{2.6}$$

$$= \mathbf{u}(\,|\,f - g\,|\,).$$

And similarly for $|\,\boldsymbol{\ell}(f) - \boldsymbol{\ell}(g)\,|\,.$

QED

For $f \geq 0$, since $\|f\| \cdot \mathbf{1} \in U$, we have $f \leq u(f) \leq \|f\| \cdot \mathbf{1}$, whence $\|f\| = \|u(f)\|$. It follows that for every $f \in C''$,

$$(*) \quad \|f\| = \| |f| \| = \|u(|f|)\|.$$

Combining this with (2.11), we obtain that for every $f \in C''$,

$$\|f\| = \max(\|u(f)\|, \|\ell(f)\|).$$

(2.13) For $f, g \in C''$,

$$\begin{array}{c} \|u(f) - u(g)\| \\[2ex] \qquad \leq \|f - g\|. \\[2ex] \|\ell(f) - \ell(g)\| \end{array}$$

Thus the operations $u(\cdot)$ and $\ell(\cdot)$ are norm continuous.

This follows from (2.12) and the above discussion.

The following will be important.

(2.14) For every $f \in C''$,

$$\ell(f)_a = f_a = u(f)_a.$$

Proof. We prove the second equality. Consider $f \in C''$, and suppose $f_a \neq u(f)_a$. So there exists $x \in X$ such that $\langle f_a, x \rangle < \langle u(f)_a, x \rangle$, that is,

$$\langle u(f)_a, x \rangle = \langle f_a, x \rangle + \lambda, \quad \lambda > 0.$$

Then

$$u(f)_a \geq f_a + \lambda \xi_x,$$

where ξ_x is the characteristic function of x. It follows

(i) $\mathbf{u}(f) - \lambda \xi_x \geq f.$

In effect,

$$(\mathbf{u}(f) - \lambda \xi_x)_a = \mathbf{u}(f)_a - \lambda \xi_x \geq f_a,$$

$$(\mathbf{u}(f) - \lambda \xi_x)_d = \mathbf{u}(f)_d \geq f_d,$$

which gives us (i). Now $\lambda \xi_x \in U$ [3; (54.13)], so by the discussion following (2.2), the left side of (i) is a $\mathbf{u}$-element. This contradicts the definition of $\mathbf{u}(f)$.

QED

An immediate consequence (remember, U itself is, in general, not a Riesz ideal):

(2.15) (1) $U \cap C_a''$ is a Riesz ideal of C''.

(2) It is the largest Riesz ideal of C'' contained in U.

Proof. (1) We need only show that if $0 \leq g \leq f \in U \cap C_a''$, then $g \in U$. By (2.2), $g \in U^\ell$; we show that $g \in U^u$ also, hence $g \in U$. By hypothesis, $0 \leq g \leq f \in U$. It follows $0 \leq \mathbf{u}(g) \leq f$, hence $\mathbf{u}(g) \in C_a''$. (2.14) above then gives us $g = \mathbf{u}(g)$.

(2) Consider a Riesz ideal G of C'' contained in U. It suffices to show $G \subset C_a''$. Now, $G_d = G \cap C_d'' \subset U \cap C_d'' = 0$ (cf. Property (U5) at the end of §1). Thus, $G = G_a \subset C_a''$.

QED

§3. The "boundary" of an element

As in [3], for each $f \in C''$, we define

$$\delta(f) = \mathbf{u}(f) - \boldsymbol{\ell}(f).$$

It is immediate that $\delta(f) \geq 0$ and $\delta(f) = 0$ if and only if $f \in U$. Also that $\delta(f)$ is a $\mathbf{u}$-element.

(2.3) gives us

(3.1) $\delta(\rho f) = |\rho|\, \delta(f)$ for all $f \in C''$ and $\rho \in \mathbb{R}$.

And (2.4) gives us

(3.2)   For $f, g \in C''$,

$$\delta(f) - \delta(g) \;\leq\; \begin{matrix} \delta(f+g) \\[4pt] \\[4pt] \delta(f-g) \end{matrix} \;\leq\; \delta(f) + \delta(g).$$

Interchanging f and g in this, and using $\delta(-f) = \delta(f)$, we obtain

(3.3)   For every $f, g \in C''$,

$$|\,\delta(f) - \delta(g)\,| \;\leq\; \begin{matrix} \delta(f+g). \\[4pt] \\[4pt] \delta(f-g). \end{matrix}$$

From (2.8), we have

(3.4)   If $g \in U$, then for every $f \in C''$,

$$\delta(f+g) = \delta(f).$$

We sharpen one of the inequalities in (3.2).

(3.5)   For $f, g \in C''$,

$$\delta(f+g) \leq \delta(f \vee g) + \delta(f \wedge g) \leq \delta(f) + \delta(g).$$

Proof.

$$\delta(f+g) = \delta(f \vee g + f \wedge g)$$

$$\leq \delta(f \vee g) + \delta(f \wedge g) \qquad\qquad (3.2)$$

$$= u(f \vee g) - \ell(f \vee g) + u(f \wedge g) - \ell(f \wedge g)$$

$$= u(f) \vee u(g) - \ell(f \vee g) + u(f \wedge g) - \ell(f) \wedge \ell(g) \qquad (2.6)$$

$$\leq u(f) \vee u(g) - \ell(f) \vee \ell(g) + u(f) \wedge u(g) - \ell(f) \wedge \ell(g) \qquad (2.5)$$

$$= [u(f) \vee u(g) + u(f) \wedge u(g)] - [\ell(f) \vee \ell(g) + \ell(f) \wedge \ell(g)]$$

$$= [u(f) + u(g)] - [\ell(f) + \ell(g)]$$

$$= [u(f) - \ell(f)] + [u(g) - \ell(g)]$$

$$= \delta(f) + \delta(g). \qquad\qquad\qquad\qquad \text{QED}$$

Setting $g = 0$, we obtain

(3.6)   For every $f \in C''$,

$$\delta(f) = \delta(f^+) + \delta(f^-).$$

It follows from this (and (3.5)) that $\delta(\,|\,f\,|\,) \leq \delta(f)$. Note also that (2.13) gives us

$$\|\,\delta(f) - \delta(g)\,\| \;\leq\; 2\,\|\,f - g\,\|\,,$$

and thus the operation $\delta(\,\cdot\,)$ is norm continuous.

By the discussion preceding (2.5), if a set $\{f_\alpha\}$ is bounded, then so is $\{\delta(f_\alpha)\}$.

(3.7)   For a countable bounded set $\{f_n\}$ in C'',

$$\begin{array}{c} \delta(\vee_n f_n) \\[2mm] \qquad\qquad \leq \; \vee\,\delta(f_n). \\[2mm] \delta(\wedge_n f_n) \end{array}$$

The verification is straightforward (using (2.5) and (2.6)).

We have pointed out that for every $f \in C''$, $\delta(f) \in U^u$. We show now that $\delta(f) \in C_d''$, so $\delta(f) \in U^u \cap C_d''$. even more:

(3.8) $\delta(C'') = U^u \cap (C_d'')_+.$

Proof. By (2.14), for every $f \in C''$, $\delta(f)_a = 0$, hence $\delta(f) \in C_d''$ (so $\delta(f) \in (C_d'')_+$). Conversely, suppose $g \in U^u \cap (C_d'')_+$. Then $g = \delta(g)$: in effect, by (2) of (2.2), $\ell(g) = 0$, hence

$$\delta(g) \;=\; u(g) - \ell(g) \;=\; u(g) \;=\; g$$

(the last equality since $g \in U^u$).

$$\text{QED}$$

(3.9) *Corollary.* The operation $\delta(\cdot)$ is idempotent.

Proof.   Given $f \in C''$, $\delta(f) \in U^u \cap (C''_d)_+$,   so (cf. the above proof), $\delta(\delta(f)) = \delta(f)$.

QED

We close this § with an important property of $\delta(\cdot)$.

(3.10) Let H be a Riesz ideal of C''.

    (I) If H is norm closed, then $\delta^{-1}(H)$

        (1) is a norm closed Riesz subspace of C'',

        (2) contains U (in particular **1**),

        (3) is closed under the operations $u(\cdot)$, $\ell(\cdot)$, and $\delta(\cdot)$.

    (II) If H is σ-order closed, then so is $\delta^{-1}(H)$.

    (III) If H is a band, then $\delta^{-1}(H)$ is Dedekind closed in C''.

Proof.   The proof of (I) is the same as that of [3; (50.11)], and the proof of (III) is the same as that of [3; (50.12)]. It remains to prove (II). Since $\delta^{-1}(H)$ is a Riesz subspace, it suffices to show that if $\{f_n\} \subset \delta^{-1}(H)$ and $f = \vee_n f_n$, then $f \in \delta^{-1}(H)$. $\{\delta(f_n)\}$ is bounded above (cf. the comment preceding (3.7)), so $\vee_n \delta(f_n)$ exists, and since H is σ-order closed, $\vee_n \delta(f_n) \in H$.   Now by (3.7), $0 \leq \delta(f) \leq \vee_n \delta(f_n)$, so $\delta(f) \in H$.

QED

§4. The "integrable elements"

We are now in position to define and study the elements of C'' which are "Lebesgue integrable" with respect to Radon measures. In the remainder of this paper, J will be a fixed band of C' and I will be the band of C'' dual to J. Thus $I = (J^{\perp})^d = (J^d)^{\perp}$ and $J = (I^d)_{\perp} = (I_{\perp})^d$.

We denote $J^{\perp}$ by $N(J)$, or simply N. And we denote $\delta^{-1}(N)$ by $\mathfrak{L}^{\infty}(J)$, or simply $\mathfrak{L}^{\infty}$. We will call the elements of $\mathfrak{L}^{\infty}$ J-integrable, or since J is fixed, simply integrable. We will write $\mathfrak{L}^{\infty}_a$ for $(\mathfrak{L}^{\infty})_a$ (recall that $(\mathfrak{L}^{\infty})_a$ is the image of $\mathfrak{L}^{\infty}$ under the projection of C'' onto C''_a). We will see later that $\mathfrak{L}^{\infty}_a$ is precisely the set of bounded functions on X which are integrable in the standard sense with respect to every $\nu \in J_+$. (And that for every $f \in \mathfrak{L}^{\infty}$ and $\nu \in J_+$, $\int f_a d\nu = \langle f, \nu \rangle$.)

By (3.10), we have

(4.1) $\mathfrak{L}^{\infty}$ is a Riesz subspace of C'' which:

(1) contains U,

(2) is Dedekind closed,

(3) is σ-order closed,

(4) is closed under the operations $u(\cdot)$, $\ell(\cdot)$, and $\delta(\cdot)$.

The following is immediate from the definition of $\mathfrak{L}^{\infty}$.

(4.2) For every $f \in C''$, $f \in \mathfrak{L}^\infty$ if and only if there exist an $\boldsymbol{\ell}$-element g and a **u**-element h such that:

      (1) $g \leq f \leq h$,

      (2) $h - g \in N$.

(4.3) *Corollary 1.* Given an $\boldsymbol{\ell}$-element g and a **u**-element h with $g \leq h$, if $h - g \in N$, then the interval $[g, h]$ is contained in $\mathfrak{L}^\infty$.

    In particular, for every $h \in U^u \cap N_+$, $[0, h] \subset \mathfrak{L}^\infty$.

(4.4) *Corollary 2.* Given $g, h \in \mathfrak{L}^\infty$ with $g \leq h$, if $h - g \in N$, then $[g, h] \subset \mathfrak{L}^\infty$ — indeed $[\boldsymbol{\ell}(g), \mathbf{u}(h)] \subset \mathfrak{L}^\infty$. In particular, for every $h \in \mathfrak{L}^\infty \cap N_+$, $[0, \mathbf{u}(h)] \subset \mathfrak{L}^\infty$.

    Proof. By Corollary 1, it suffices to show that $\mathbf{u}(h) - \boldsymbol{\ell}(g) \in N$. We have

$$0 \leq \mathbf{u}(h) - h \leq \boldsymbol{\delta}(h) \in N,$$
$$0 \leq g - \boldsymbol{\ell}(g) \leq \boldsymbol{\delta}(g) \in N.$$

Hence

$$0 \leq \mathbf{u}(h) - \boldsymbol{\ell}(g) = (\mathbf{u}(h) - h) + (h - g) + (g - \boldsymbol{\ell}(g)) \in N.$$

                                        QED

§5. The negligible elements

We will denote $\mathfrak{L}^{\infty} \cap N$ by $\mathcal{N}(J)$, or simply $\mathcal{N}$, and call its elements *negligible*.

(5.1)   (1) $\mathcal{N}$ is a σ-order closed Riesz ideal of C''.

(2)   It is closed under the operations $\mathbf{u}(\cdot)$, $\boldsymbol{\ell}(\cdot)$, and $\delta(\cdot)$.

(3)   It is the Riesz ideal generated by $\mathbf{u}$-elements in N_+.

Proof. (1) $\mathfrak{L}^{\infty}$ is σ-order closed and N is order closed, so $\mathcal{N}$ is σ-order closed. That $\mathcal{N}$ is a Riesz ideal follows from (4.4).

(2)   This follows from the proof of (4.4).

(3)   By the comment following (4.3), the Riesz ideal generated by $U^{\mathbf{u}} \cap N_+$ is contained in $\mathcal{N}$. Conversely, if $f \in \mathcal{N}_+$, then $0 \le f \le \mathbf{u}(f)$, and by (2), $\mathbf{u}(f) \in \mathcal{N} \subset N$.

                                                                QED

In particular, $U^{\mathbf{u}} \cap N_+ \subset \mathcal{N}$, so for every $f \in \mathfrak{L}^{\infty}$, $\delta(f) \in \mathcal{N}$.

$\mathcal{N}$ can be defined directly, without first defining $\mathfrak{L}$. Its defining condition has, of course, to be stronger than that for $\mathfrak{L}$. While $f \in \mathfrak{L}$ is defined by $\mathbf{u}(f) - \boldsymbol{\ell}(f) \in N$, we have

(5.2)   For every $f \in C''$, $f \in \mathcal{N}$ if and only if $\mathbf{u}(f)$ and $\boldsymbol{\ell}(f)$ both lie in N.

Since every $f \in C''$ can be written $f = \boldsymbol{\ell}(f) + (f - \boldsymbol{\ell}(f)) = \mathbf{u}(f) - (\mathbf{u}(f) - f)$, we have

(5.3) Every $f \in \mathfrak{L}^{\infty}$ can be written

$$f = g + p, \quad \text{g an } \boldsymbol{\ell}\text{-element and } p \in \mathcal{N}$$

$$= h + g, \quad \text{h a u-element and } g \in \mathcal{N}.$$

(5.4) $\mathcal{N}$ is order dense in N.

Remark. Since $\mathcal{N}$ is a Riesz ideal, the above means that every $f \in N_+$ is the supremum of some subset of $\mathcal{N}_+$.

Proof of (5.4). Recall that for each $\mu \in C'_+$, C'_μ is the band of C' generated by μ and C''_μ the band of C'' dual to C'_μ. We showed in [3; §41] that $\Sigma_{\mu \in C'_+} C''_\mu$ is a σ-order closed Riesz ideal order dense in C''. Hence to prove (5.4), it suffices to show that $(\Sigma_{\mu \in C'_+} C''_\mu) \cap N \subset \mathcal{N}$ — equivalently, for every $\mu \in C'_+$, $C''_\mu \cap N \subset \mathcal{N}$. Consider $f \in C''_\mu \cap N$, and we can take $f \geq 0$. By (2.2), $f \in U^u$, hence by (5.1), $f \in \mathcal{N}$.

QED

The following should be compared with (2.15). (It actually contains (2.15); cf. the examples at the end of §6.)

(5.5)   (1) $\mathfrak{L}^{\infty} \cap C''_a$ is a Riesz ideal of C''.

   (2) $\mathfrak{L}^{\infty} \cap C''_d$ is a Riesz ideal of C''.

   (3) $(\mathfrak{L}^{\infty} \cap C''_a) \oplus (\mathfrak{L}^{\infty} \cap C''_d)$ is the largest Riesz ideal of C'' contained in $\mathfrak{L}^{\infty}$.

Proof. (1) It suffices to show that $0 \leq g \leq f \in \mathfrak{L}^\infty \cap C_a''$ implies $g \in \mathfrak{L}^\infty$; and for this, it is enough to show that $\delta(g) \leq \delta(f)$. g and f are ℓ-elements (cf. (2.2)), so $\delta(g) = u(g) - g$ and $\delta(f) = u(f) - f$. Hence, by (3.8),

$$\delta(g) = \delta(g)_d = u(g)_d - g_d = u(g)_d;$$

and similarly, $\delta(f) = u(f)_d$.   Now $g \leq f$, so $u(g) \leq u(f)$, so $u(g)_d \leq u(f)_d$. Thus $\delta(g) \leq \delta(f)$.

(2) It suffices to show that $\mathfrak{L}^\infty \cap C_d'' = N \cap C_d''$; and for this, it is enough to show that $\mathfrak{L}^\infty \cap C_d'' \subset N$. Consider $f \in \mathfrak{L}^\infty \cap C_d''$, and we can take $f \geq 0$.   By (2.2), $\ell(f) = 0$, so $u(f) = \delta(f)$, so $0 \leq f \leq \delta(f) \in N$, whence $f \in N$.

(3) For every Riesz ideal G of C'' contained in $\mathfrak{L}^\infty$, $G = (G \cap C_a'') \oplus (G \cap C_d'') \subset (\mathfrak{L}^\infty \cap C_a'') \oplus (\mathfrak{L}^\infty \cap C_d'')$.

QED

§6. The "μ-integrable elements"

In this §, we assume $J = C_\mu'$ for a fixed $\mu \in C'$, $\mu \geq 0$. We will denote N, $\mathfrak{L}^\infty$, and N by $N(\mu)$, $\mathfrak{L}^\infty(\mu)$, and $N(\mu)$ respectively.

Note first that $N(\mu)$ can be described in terms of μ alone (instead of all of C_μ'): $f \in N(\mu)$ if and only if $\langle |f|, \mu \rangle = 0$. (This is well known, but we give the details in (10.2).) This in turn gives us a simple characterization of $\mathfrak{L}^\infty(\mu)$:

(6.1) For $f \in C''$, the following are equivalent:

$1°\ f \in \mathfrak{L}^{\infty}(\mu),$

$2°\ \langle \boldsymbol{\ell}(f), \mu \rangle = \langle f, \mu \rangle = \langle \mathbf{u}(f), \mu \rangle.$

By "f is μ-integrable", we will mean 2° above. Thus (6.1) says that for $J = C'_{\mu}$ $(\mu \geq 0)$, f is J-integrable if and only if it is μ-integrable.

(4.2) takes, in the present case, the following stronger form ((6.2) and (6.3)):

(6.2) For $f \in C''$, the following are equivalent:

$\quad$ $1°\ f \in \mathfrak{L}^{\infty}(\mu);$

$\quad$ $2°$ there exist $g, h \in U$ such that

$\quad\quad$ (1) $g \leq f \leq h,$

$\quad\quad$ (2) $h - g \in N(\mu).$

Proof. For $k \in U$, $\boldsymbol{\ell}(k) = k = \mathbf{u}(k)$, so we need only show that $1°$ implies $2°$. Assume $1°$.

Lemma. There exist $g, h \in U$ such that

$$g \leq \boldsymbol{\ell}(f) \leq f \leq \mathbf{u}(f) \leq h,$$

$$\langle g, \mu \rangle = \langle \boldsymbol{\ell}(f), \mu \rangle,$$

$$\langle h, \mu \rangle = \langle \mathbf{u}(f), \mu \rangle.$$

μ is order continuous on C'', so (by the definition of U^{ℓ} and U^{u}), for each $\epsilon > 0$, there exists $g(\epsilon) \in U$ such that

$$\langle g(\epsilon), \mu \rangle \geq \langle \ell(f), \mu \rangle - \epsilon.$$

It follows we can choose $\{g_n\} \subset U$ satisfying

$$g_1 \leq g_2 \leq \cdots \leq \ell(f),$$

$$\langle g_n, \mu \rangle \geq \langle \ell(f), \mu \rangle - 1/n \quad (n=1,2,...).$$

Since U is σ-order closed, $\vee_n g_n \in U$ and $\vee_n g_n$ is the desired g. h is obtained similarly.

This establishes the Lemma. Combining it with (6.1) gives us $2°$.   QED

By (2.3), the g_n's in the proof of the above Lemma can be chosen to be usc elements, and the h_n's used to obtain the desired h can be chosen to be lsc elements. So (6.2) can be put in the form

(6.3) For $f \in C''$, the following are equivalent:

   $1°$ $f \in \mathfrak{L}^\infty(\mu)$;

   $2°$ there exist sequences $\{g_n\}$, $\{h_n\}$ of usc elements and lsc elements respectively such that

$$g_1 \leq g_2 \leq \cdots \leq f \leq \cdots \leq h_2 \leq h_1,$$

$$\sup_n \langle g_n, \mu \rangle = \langle f, \mu \rangle = \inf_n \langle h_n, \mu \rangle.$$

In the present case, we can also sharpen the results on $\mathcal{N}$. For example, (3) in (5.1) takes the form: $\mathcal{N}(\mu)$ is the Riesz ideal of C'' generated by $U \cap N(\mu)_+$ (by the argument used for the Lemma above). And (5.3) takes the form

(6.4) $\mathfrak{L}^\infty(\mu) = U + \mathcal{N}(\mu)$

$\qquad = \mathfrak{B}o + \mathcal{N}(\mu)$

$\qquad = \mathfrak{B}a + \mathcal{N}(\mu)$

$\qquad = \mathfrak{B}a_2 + \mathcal{N}(\mu).$

Proof. $\mathfrak{B}a_2 + \mathcal{N}(\mu) \subset \mathfrak{B}a + \mathcal{N}(\mu) \subset \mathfrak{B}o + \mathcal{N}(\mu) \subset U + \mathcal{N}(\mu) \subset \mathfrak{L}^\infty(\mu).$ So we need only show that $\mathfrak{L}^\infty(\mu) \subset \mathfrak{B}a_2 + \mathcal{N}(\mu).$ Consider $f \in \mathfrak{L}^\infty(\mu).$ By [2; (41.3)], there exists $g \in \mathfrak{B}a_2$ such that $g_\mu = f_\mu.$ Thus $f - g \in N(\mu),$ and since $f - g \in \mathfrak{L}^\infty(\mu),$ we have $f - g \in \mathcal{N}(\mu).$ So $f = g + (f - g) \in \mathfrak{B}a_2 + \mathcal{N}(\mu).$

$\qquad\qquad\qquad\qquad\qquad\qquad\qquad\qquad\qquad$ QED

Now drop the assumption that $J = C''_\mu$ for some $\mu.$

(6.5)   (1) $N = \cap_{\nu \in J_+} N(\nu).$

$\qquad$ (2) $\mathfrak{L}^\infty = \cap_{\nu \in J_+} \mathfrak{L}^\infty(\nu).$

$\qquad$ (3) $\mathcal{N} = \cap_{\nu \in J_+} \mathcal{N}(\nu).$

Thus $f \in C''$ is J-integrable if and only if it is ν-integrable for every $\nu \in J_+.$

We close with some elementary examples.

(I) Let $J = C'.$ Then $N = 0,$ so $\mathfrak{L}^\infty = U$ and $\mathcal{N} = 0.$

(II) Let $J = C'_d.$ Then $N = C''_a,$ so (by (3.8)) $\mathfrak{L}^\infty = U$ and $\mathcal{N} = U \cap C''_a.$

(III) Let $J = C'_a$. Then $N = C''_d$, so (again by (3.8)) $\mathfrak{L}^\infty = C''$ and $\mathcal{N} = N = C''_d$.

(IV) Let $J = 0$. then $\mathfrak{L}^\infty = N = \mathcal{N} = C''$.

By (I), U is the set of elements of C'' μ-integrable for every Radon measure μ. And by (II), it is also the set of elements of C'' μ-integrable for every diffuse Radon measure μ. This is because, by (III), every element of C'' is μ-integrable for every atomic Radon measure μ.

Chapter 2. Convergence

The gain in richness of Lebesgue over Riemann integration is due to the convergence properties of the former. In this chapter we discuss various forms of convergence which play a role in the theory. We will confine ourselves to convergence of sequences. The reason is that $\mathfrak{L}^\infty$ and $\mathcal{N}$ are σ-order closed but, in general, not order closed. (We remark here that the σ-order closedness of $\mathfrak{L}^\infty$ is not part of its definition and is not based on any countability assumption. It follows from the basic result (2.1).)

§7. Almost everywhere order convergence

The following property of $\mathcal{N}$ is easily verified. It is important to remember here, and in the "almost everywhere" definitions below, that $\mathcal{N}$ itself is, in general, not a band of C''.

(7.1) For every countable family $\{H_n\}$ of bands of C'' contained in $\mathcal{N}$, the band of C'' generated by $\cup_n H_n$ is also contained in $\mathcal{N}$.

Given $f, g \in C''$, then by $g = f$ J-*almost everywhere*, or J-*a.e.*, we will mean there exists a band H of C'' contained in $\mathcal{N}$ such that $g_H d = f_H d$. However, since J is fixed, we will simply write $g = f$ a.e., omitting J.

31

(7.2)   (1)   g = f implies g = f a.e..

      (2)   g = f a.e. is equivalent to f = g a.e..

      (3)   g = f a.e. and f = h a.e. imply g = h a.e..

      (4)   g = f a.e. implies $\rho g = \rho f$ a.e. for every $\rho \in \mathbb{R}$.

      (5)   If g = f a.e. and k = h a.e., then:

         (i)   g + k = f + h a.e.,

         (ii)   g ∨ k = f ∨ k a.e.,

         (iii)   g ∧ k = f ∧ k a.e..

The verifications are straightforward.

(7.3)   *Corollary.* For $f, g \in C''$, the following are equivalent:

      1°   g = f a.e.;

      2°   (i) $g^+ = f^+$ a.e.,

         (ii) $g^- = f^-$ a.e..

And they imply

$$|g| = |f| \text{ a.e..}$$

Given $f, g \in C''$, then by $g \leq f$ J-*a.e.*, we will mean that there exists a band H of C'' contained in N such that $g_{H^d} \leq f_{H^d}$ (equivalently, $(g-f)^+ = 0$ a.e.). Again, since J is fixed, we will simply write $g \leq f$ a.e.

We will feel free to write the sequence $g \leq f$ a.e., $f \leq h$ a.e. in the form $g \leq f \leq h$ a.e..

(7.4) (1) $g \leq f$ implies $g \leq f$ a.e. .

(2) $g \leq f$ a.e. and $f \leq g$ a.e. imply $g = f$ a.e. .

(3) $g \leq f$ a.e. and $f \leq h$ a.e. imply $g \leq h$ a.e. .

(4) $g \leq f$ a.e. implies $\rho g \leq \rho f$ a.e. for every $\rho \geq 0$.

(5) If $g \leq f$ a.e. and $k \leq h$ a.e., then:

(i) $g + k \leq f + h$ a.e.,

(ii) $g \vee k \leq f \vee h$ a.e.,

(iii) $g \wedge k \leq f \wedge h$ a.e. .

The verifications are straightforward.

(7.5) *Corollary 1.* For $f, g \in C''$, the following are equivalent:

$1°$ $g \leq f$ a.e.;

$2°$ (i) $g^{+} \leq f^{+}$ a.e.,

(ii) $f^{-} \leq g^{-}$ a.e. .

(7.6) *Corollary 2.* (1) $g, h \leq f$ a.e. implies $g \vee h \leq f$ a.e. .

(2) The following are equivalent:

$1°$ $g \leq f$ a.e.,

$2°$ $g = f \wedge g$ a.e. .

(3) The following are equivalent:

$1°$ $g \leq f$ a.e.,

$2°$ $f - g \geq 0$ a.e. .

(7.7)   If $C'' = G_1 \oplus G_2$, G_1 and G_2 bands, then for $f, g \in C''$, the following are equivalent:

    $1°$ $g \leq f$ a.e.;

    $2°$ (i) $g_{G_1} \leq f_{G_1}$ a.e.,

        (ii) $g_{G_2} \leq f_{G_2}$ a.e..

It follows the equivalence holds with $\leq$ replaced by $=$.

Proof. Assume $1°$. So there exists a band H of C'' contained in $\mathcal{N}$ such that $g_{H^d} \leq f_{H^d}$. Then

$$(g_{G_1})_{H^d} = (g_{H^d})_{G_1} \leq (f_{H^d})_{G_1} = (f_{G_1})_{H^d}.$$

Thus $g_{G_1} \leq f_{G_1}$ a.e.. Similarly, $g_{G_2} \leq f_{G_2}$ a.e.. That $2°$ implies $1°$ follows from (7.4).

QED

Given $A \subset C''$ and $h \in C''$, then by $A \leq h$ J-*a.e.*, we will mean that there exists a band H of C'' contained in $\mathcal{N}$ such that $A_{H^d} \leq h_{H^d}$. Since J is fixed, we will simply write $A \leq h$ a.e.. we will say that A is *bounded above a.e.* by h.

Note that our definition is stronger than the statement "$f \leq h$ a.e. for every $f \in A$." Distinct f's will, in general, require different H's, while our definition postulates a single H for all the elements of A. However,

(7.8) For a countable set $A = \{f_n\}$, the following are equivalent:

      1° $\{f_n\} \leq h$ a.e.,

      2° $f_n \leq h$ a.e. for every n.

Proof. Assume 2°. So for each n, there exists a band H_n of C'' contained in $\mathcal{N}$ such that $(f_n)_{(H_n)^d} \leq h_{(H_n)^d}$. We have to find a single band H of C'' contained in $\mathcal{N}$ such that

$$(f_n)_{H^d} \leq h_{H^d} \quad \text{for every n.}$$

The band H generated by $\cup_n H_n$ has this property (cf. (7.1)).

QED

Bounded below a.e. is defined similarly, and (7.8) holds with $\leq$ replaced by $\geq$. We will say A is *bounded J-a.e.*, or simply *a.e.*, if there exists a band H of C'' contained in $\mathcal{N}$ such that A_{H^d} is bounded. It is easily seen that this is equivalent to the statement that A is bounded above a.e. and bounded below a.e..

Boundedness implies boundedness a.e.. The converse is false. However if a set A is (say) bounded above a.e., then by definition, A_{H^d} is bounded above for an appropriate band H. This will enable us to reduce a question involving boundedness above a.e. to one involving ordinary boundedness above. And similarly for boundedness below a.e. and boundedness a.e..

There are two formally different definitions for "$h = \vee A$ a.e.". We first show they are equivalent.

(7.9) For $A \subset C''$ and $h \in C''$, the following are equivalent:

1° there exists a band H of C'' contained in $\mathcal{N}$ such that

$$h_{H^d} = \vee A_{H^d};$$

2° (i) A is bounded above a.e. by h,

(ii) for every $k \in C''$, if A is bounded above a.e. by k, then

$k \geq h$ a.e..

Proof. Assume 1°. $h_{H^d} \geq A_{H^d}$, so (i) holds. Suppose $k \geq A$ a.e.. This says there exists a band H_1 of C'' contained in $\mathcal{N}$ such that $k_{(H_1)^d} \geq A_{(H_1)^d}$. Set $H_2 = H + H_1$. Then $h_{(H_2)^d} = \vee A_{(H_2)^d}$ and $k_{(H_2)^d} \geq A_{(H_2)^d}$, hence $k_{(H_2)^d} \geq h_{(H_2)^d}$. Thus (ii) holds.

Assume 2°. By (i), there exists a band H_1 of C'' contained in $\mathcal{N}$ such that $A_{(H_1)^d} \leq h_{(H_1)^d}$. Set $f = \vee A_{(H_1)^d}$. We note first that

(∗) $h = f$ a.e..

In effect: On the one hand, by the definition of f, $f \leq h_{(H_1)^d}$; hence, since $f = f_{(H_1)^d}$, $f \leq h$ a.e.. On the other hand, $f \geq A_{(H_1)^d}$, so (again since $f = f_{(H_1)^d}$) $f \geq A$ a.e.; hence by (ii), $f \geq h$ a.e.. Thus $f = h$ a.e..

(∗) says there exists a band H_2 of C'' contained in $\mathcal{N}$ such that $h_{(H_2)^d} = f_{(H_2)^d}$. Set $H = H_1 + H_2$. then $h_{H^d} = f_{H^d} = \vee A_{H^d}$, so 1° holds.

QED

If A and h satisfy one, hence also the other, of the above conditions, we will write $h = \vee A$ *J-a.e.*, or simply $h = \vee A$ *a.e.*.

"$h = \wedge A$ a.e." is defined similarly. In the following, propositions stated for $h = \vee A$ a.e. will, without specific mention, hold also for $h = \wedge A$ a.e.

The verifications of the following are straightforward.

(7.10) (1) If $h = \vee_n f_n$ a.e., then $\rho h = \vee_n (\rho f_n)$ a.e. for every $\rho \geq 0$.

(2) If $h = \vee_n f_n$ a.e., then $-h = \wedge_n (-f_n)$ a.e. .

(3) If $h = \vee_n f_n$ a.e. and $k = h$ a.e., then $k = \vee_n f_n$ a.e. .

(4) If $h = \vee_n f_n$ a.e. and $k = \vee_n f_n$ a.e., then $k = h$ a.e. .

(5) If $h = \vee_n f_n$ a.e. and $g_n = f_n$ a.e. $(n = 1, 2, ...)$, then $h = \vee_n g_n$ a.e. .

(6) If $h = \vee_n f_n$ a.e., $k = \vee_n g_n$ a.e., and $g_n \leq f_n$ a.e. for every n, then $k \leq h$ a.e. .

(7) If $h = \vee_n f_n$ a.e. and $k = \vee_n g_n$ a.e., then:

(i) $h \vee k = \vee_n (f_n \vee g_n)$ a.e.

(ii) $h \wedge k = \vee_{n,m} (f_n \wedge g_m)$ a.e.,

(iii) $h + k = \vee_{n,m} (f_n + g_m)$ a.e. .

Remark. As is easily verified from the above, the following are equivalent:

$$1° \quad h = \vee_n f_n \text{ a.e.,}$$

$$2° \quad 0 = \wedge_n (h - f_n) \text{ a.e. .}$$

(7.11)   Suppose $C'' = G_1 \oplus G_2$, G_1 and G_2 bands.   Then for $\{f_n\} \subset C''$ and $h \in C''$, the following are equivalent:

$1°\ \ h = \vee_n f_n$ a.e.;

$2°$ (i) $h_{G_1} = \vee_n (f_n)_{G_1}$ a.e.,

    (ii) $h_{G_2} = \vee_n (f_n)_{G_2}$ a.e. .

We list some obvious properties which we will be using in the paper. $h = \vee_n f_n$ implies $h = \vee_n f_n$ a.e. . So given $f \in C''$, then for every $g \in C''_+$, $g_f = \vee_n (g \wedge n |f|)$ a.e. . It follows that if $h = k$ a.e., then $\mathbf{1}_h = \mathbf{1}_k$ a.e. . Note also that $\mathbf{1}$ is a "weak order unit a.e.": $f \wedge \mathbf{1} = 0$ a.e. implies $f = 0$ a.e. .

A sequence $\{f_n\}$ in C'' will be said to be *ascending J-a.e.*, or simply *a.e.* if there exists a band H of C'' contained in $\mathcal{N}$ such that $\{(f_n)_{H^d}\}$ is ascending, that is, $n \leq m$ implies $(f_n)_{H^d} \leq (f_m)_{H^d}$. It is easy to verify, using (7.1), that an equivalent definition is:

$n \leq m$ implies $f_n \leq f_m$ a.e. .

Descending a.e. is defined similarly.

Remark. The above definitions are *not* equivalent for a general net (the first is stronger). It follows this remark holds also for the definitions below of order convergence a.e. .

We sharpen (7) in (7.10).

(7.12) Given $h = \vee_n f_n$ a.e. and $k = \vee_n g_n$ a.e., if $\{h_n\}$ and $\{g_n\}$ are ascending a.e., then:

(2) $h \wedge k = \vee_n (f_n \wedge g_n)$ a.e.,

(3) $h + k = \vee_n (f_n + g_n)$ a.e..

As usual, the notation $f_n \uparrow h$ will mean that $\{f_n\}$ is ascending and $h = \vee_n f_n$. And similarly for $f_n \downarrow g$. Now, for a sequence $\{f_n\}$ in C'' and $h \in C''$, the following are equivalent:

1°   there exists a band H of C'' contained in $\mathcal{N}$ such that $(f_n)_{H^d} \uparrow h_{H^d}$;

2° (i) $\{f_n\}$ is ascending a.e.,

    (ii) $h = \vee_n f_n$ a.e..

We leave the verification to the reader. If one, hence also the other, of these conditions holds, we will write

$$f_n \uparrow h \quad \text{a.e..}$$

A similar discussion holds for $f_n \downarrow g$ a.e..

We can now define almost everywhere convergence. For a sequence $\{f_n\} \subset C''$ and $f \in C''$, the following are equivalent:

1°   there exists a band H of C'' contained in $\mathcal{N}$ such that $(f_n)_{H^d} \to f_{H^d}$;

2° there exist sequences $\{g_n\}$, $\{h_n\}$ in C'' such that

    (i) $h_n \downarrow f$ a.e.,

    (ii) $g_n \uparrow f$ a.e.,

    (iii) $g_n \le f_n \le h_n$ a.e. for every n.

We leave the verification to the reader. If one, hence also the other, of these holds, we will say that $\{f_n\}$ order converges to f J-*almost everywhere*, or simply *almost everywhere* — in notation, $f_n \longrightarrow f$ a.e.

The notation $(f_n)_{H^d} \longrightarrow f_{H^d}$ carries with it, as always, the understanding that $\{(f_n)_{H^d}\}$ is bounded. It follows the notation $f_n \longrightarrow f$ a.e. carries with it the understanding that $\{f_n\}$ is bounded a.e..

(7.13) (1) If $f_n \longrightarrow f$ a.e., then $\rho f_n \longrightarrow \rho f$ a.e. for every $\rho \in \mathbb{R}$.

(2) If $f_n \longrightarrow f$ a.e. and $g = f$ a.e., then $f_n \longrightarrow g$ a.e..

(3) If $f_n \longrightarrow f$ a.e. and $g_n = f_n$ a.e. $(n = 1, 2, ...)$, then $g_n \longrightarrow f$ a.e..

(4) If $f_n \longrightarrow f$ a.e., $g_n \longrightarrow g$ a.e., and $g_n \leq f_n$ a.e. for every n, then $g \leq f$ a.e..

(5) If $g_n \longrightarrow f$ a.e., $h_n \longrightarrow f$ a.e., and $g_n \leq f_n \leq h_n$ a.e. $(n=1, 2, ...)$, then $f_n \longrightarrow f$ a.e..

(6) If $f_n \longrightarrow f$ a.e. and $g_n \longrightarrow g$ a.e., then:

(i) $f_n \vee g_n \longrightarrow f \vee g$ a.e.,

(ii) $f_n \wedge g_n \longrightarrow f \wedge g$ a.e.,

(iii) $f_n + g_n \longrightarrow f + g$ a.e..

(7.14) *Corollary 1.* For $\{f_n\} \subset C''$ and $f \in C''$, the following are equivalent:

1° $f_n \longrightarrow f$ a.e.;

2° (i) $(f_n)^+ \longrightarrow f^+$ a.e.,

(ii) $(f_n)^- \longrightarrow f^-$ a.e..

And they imply

$$|f_n| \longrightarrow |f| \quad \text{a.e..}$$

Moreover, if $f = 0$, the last statement is equivalent to $1°$ and $2°$.

(7.15) *Corollary 2.* The following are equivalent:

$1°$ $f_n \longrightarrow f$ a.e.,

$2°$ $|f_n - f| \longrightarrow 0$ a.e..

(7.16) Suppose $C'' = G_1 \oplus G_2$, G_1 and G_2 bands. Then for $\{f_n\} \subset C''$ and $f \in C''$, the following are equivalent:

$1°$ $f_n \longrightarrow f$ a.e.;

$2°$ (i) $(f_n)_{G_1} \longrightarrow f_{G_1}$ a.e.,

(ii) $(f_n)_{G_2} \longrightarrow f_{G_2}$ a.e..

An important convergence is *convergence mod N*. We record the obvious definitions.

(I) $g = f$ *mod N* if $g - f \in N$.

(II) $g \leq f$ *mod N* if $(g-f)^+ \in N$.

(III) $h = \vee A$ *mod N* if:

(i) $h \geq f$ mod N for every $f \in A$,

(ii) $k \geq f$ mod N for every $f \in A$ implies $k \geq h$ mod N.

(IV) $f_n \longrightarrow f$ *mod N* if there exist sequences $\{g_n\}$, $\{h_n\}$ in C'' satisfying:

(i) $g_n \leq g_{n+1}$ mod $\mathcal{N}$ for every n,

(ii) $h_n \geq h_{n+1}$ mod $\mathcal{N}$ for every n,

(iii) $f = \vee_n g_n$ mod $\mathcal{N} = \wedge_n h_n$ mod $\mathcal{N}$,

(iv) for every n, $g_n \leq f_n \leq h_n$ mod $\mathcal{N}$.

However, we will not concern ourselves with convergence mod $\mathcal{N}$ because of the

SCHOLIUM. For countable sets and for sequences, every J-a.e. statement is equivalent to the statement obtained by replacing "a.e." with "mod $\mathcal{N}$". (For example, $f = g$ a.e. if and only if $f - g \in \mathcal{N}$.)

We emphasize that the Scholium does not hold for uncountable sets or for general nets.

On notation. Suppose J is the band generated by a single μ — in our notation, $J = C'_\mu$ — and we can take $\mu \geq 0$. then, as we have noted in §6, N, hence also $\mathcal{N}$ and $\mathfrak{L}$, are determined by μ alone, that is, we don't have to know all of C'_μ. Consequently, for this case, we will use the notation "μ-a.e." instead of "C'_μ-a.e." or "a.e." for the almost everywhere statements.

§8. Unbounded order convergence

In this §, we discuss the extension of the definition of order convergence to sequences which are not necessarily bounded (due, to the best of our knowledge, to H. Nakano [5]). Following DeMarr [2], we will call this convergence *unbounded* order convergence. In [4], we established various characterizations of unbounded order convergence in a Dedekind complete Riesz space with a weak order unit. In this §, we record those results in [4] which we will need. The reader is referred to [4] for the proofs.

Let E be a Dedekind complete Riesz space with elements $a, b, c, \ldots$ and a weak order unit $\mathbf{1}$. A sequence $\{a_n\}$ in E will be said to order converge *unboundedly* to a — in notation, $a_n \xrightarrow{u} a$ — if for every $b, c \in E$ with $b \le c$,

$$(*) \qquad (a_n \wedge c) \vee b \longrightarrow (a \wedge c) \vee b.$$

(The definition of course holds in any Riesz space.) Note that for an order bounded sequence, this reduces to order convergence. Also, it is easy to show that it suffices for $(*)$ to hold for all b, c such that $b \le 0 \le c$. Using this, we have

(8.1) Given a band G of E, then for $\{a_n\} \subset G$ and $a \in G$, the following are equivalent:

$1°$ $a_n \xrightarrow{u} a$ in E,

$2°$ $a_n \xrightarrow{u} a$ in G considered as a Riesz space in its own right.

Two of the characterizations of unbounded order convergence obtained in [4]:

(8.2) For $\{a_n\} \subset E$ and $a \in E$, the following are equivalent:

$1°$ $a_n \xrightarrow{u} a$;

$2°$ for every $m = 1, 2, ...,$ $(a_n \wedge m\mathbf{1}) \longrightarrow (a \wedge m\mathbf{1}) \vee (-m\mathbf{1})$;

$3°$ $|a_n - a| \wedge \mathbf{1} \longrightarrow 0.$

In particular, for $\{a_n\} \subset E_+$, $a_n \xrightarrow{u} 0$ if and only if $a_n \wedge \mathbf{1} \longrightarrow 0$.

(8.3)   (1) If $a_n \xrightarrow{u} a$, then $\rho a_n \longrightarrow \rho a$ for every $\rho \in \mathbb{R}$.

(2) If $a_n \xrightarrow{u} a$ and $a_n \xrightarrow{u} b$, then $a = b$.

(3). If $a_n \xrightarrow{u} a$ and $b_n \xrightarrow{u} b$, then $a_n \vee b_n \xrightarrow{u} a \vee b$, $a_n \wedge b_n \xrightarrow{u} a \wedge b$, and $a_n + b_n \xrightarrow{u} a + b$.

(4) If $a_n \xrightarrow{u} a$, $b_n \xrightarrow{u} b$, and $b_n \leq a_n$ for every n, then $b \leq a$.

(5) If $b_n \xrightarrow{u} a$, $c_n \xrightarrow{u} a$, and $b_n \leq a_n \leq c_n$ for $n = 1, 2, ...$, then $a_n \xrightarrow{u} a$.

Remark. contained in the above is the equivalence of the following:

$1°$ $a_n \xrightarrow{u} a$,

$2°$ $a_n - a \xrightarrow{u} 0$.

And also

(8.4) The following are equivalent:

$1°$ $a_n \xrightarrow{u} a$;

$2°$ (i) $(a_n)^+ \xrightarrow{u} a^+$,

(ii) $(a_n)^- \xrightarrow{u} a^-$.

And they imply

$$|a_n| \xrightarrow{u} |a|.$$

Moreover, if $a = 0$, then the last statement is equivalent to $1°$ and $2°$.

(8.5) If $E = G_1 \oplus G_2$, G_1 and G_2 bands, then for a sequence $\{a_n\}$ in E and $a \in E$, the following are equivalent:

$1°$ $a_n \xrightarrow{u} a$;

$2°$ (i) $(a_n)_{G_1} \xrightarrow{u} a_{G_1}$,

(ii) $(a_n)_{G_2} \xrightarrow{u} a_{G_2}$.

In particular, projection upon a band preserves unbounded order convergence.

We give an additional characterization of unbounded order convergence in E. Given $a \in E_+$, then for each $\lambda \in \mathbb{R}$, we have $\mathbf{1}_{(a-\lambda\mathbf{1})^+}$, the component of $\mathbf{1}$ in the band generated by $(a - \lambda\mathbf{1})^+$. (On a set Y, if $\mathbf{1}$ is the function of constant value 1, then for a function f on Y, $\mathbf{1}_{(f-\lambda\mathbf{1})^+}$ is the characteristic function of the set $\{y \in Y : f(y) > \lambda\}$.)

(8.6) For a sequence $\{a_n\}$ in E, the following are equivalent:

$1°$ $a_n \xrightarrow{u} 0$;

$2°$ for every $\lambda > 0$, $\mathbf{1}_{(a_n - \lambda 1)^+} \longrightarrow 0$.

(Note the resemblance of the above condition to convergence in measure.)

§9. Almost everywhere unbounded order convergences

We return to C''.

(9.1) For $\{f_n\} \subset C''$ and $f \in C''$, the following are equivalent:

$1°$ there exists a band H of C'' contained in N such that

$$(f_n)_{H^d} \xrightarrow{u} f_{H^d}.$$

$2°$ for every g, h with $g \leq h$ a.e.,

$$(f_n \wedge h) \vee g \longrightarrow (f \wedge h) \vee g \text{ a.e.}.$$

Proof. Assume $1°$ and consider g, h with $g \leq h$ a.e.. Choose a band H_1 or C'' contained in N such that $g_{(H_1)^d} \leq h_{(H_1)^d}$, and set $H_2 = H + H_1$. Then

$$[(f_n \wedge h) \vee g]_{(H_2)^d} = [(f_n)_{(H_2)^d} \wedge h_{(H_2)^d}] \vee g_{(H_2)^d}$$

$$\longrightarrow [f_{(H_2)^d} \wedge h_{(H_2)^d}] \vee g_{(H_2)^d}$$

$$= [(f \wedge h) \vee g]_{(H_2)^d}.$$

Thus $(f_n \wedge h) \vee g \longrightarrow (f \wedge h) \vee g$ a.e..

Conversely, assume 2°. Then, in particular, for each $m = 1, 2, ...,$ there exists a band H_m of C'' contained in $\mathcal{N}$ such that

$$[(f_n \wedge m\mathbf{1}) \vee (-m\mathbf{1})]_{(H_m)^d} \longrightarrow [(f \wedge m\mathbf{1}) \vee (-m\mathbf{1})]_{H^d}. \qquad (8.2)$$

Let H be the band generated by $\cup_m H_m$ (cf. (7.1)). Then for every m:

$$[(f_n \wedge m\mathbf{1}) \vee (-m\mathbf{1})]_{H^d} \longrightarrow [(f \wedge m\mathbf{1}) \vee (-m\mathbf{1})]_{H^d}.$$

This can be written: for every m,

$$[(f_n)_{H^d} \wedge m\mathbf{1}_{H^d}] \vee (-m\mathbf{1}_{H^d}) \longrightarrow [f_{H^d} \wedge m\mathbf{1}_{H^d}] \vee (-m\mathbf{1}_{H^d}).$$

It follows from (8.1) and (8.2) that $(f_n)_{H^d} \xrightarrow{u} f_{H^d}$.

$$\text{QED}$$

If one, hence also the other, of the above conditions holds, we will say that $\{f_n\}$ *order converges unboundedly to* f *almost everywhere* — in notation, $f_n \xrightarrow{u} f$ *a.e.*.

Since this definition reduces almost everywhere unbounded order convergence to unbounded order convergence, it is easy to show that results in §8 carry over to the present situation. We record them.

(9.2) For $\{f_n\} \subset C''$ and $f \in C''$, the following are equivalent:

   1° $f_n \xrightarrow{u} f$ a.e.;

$2°$ for every $m = 1, 2, ...,$ $(f_n \wedge m\mathbf{1}) \vee (-m\mathbf{1}) \longrightarrow$
$(f \wedge m\mathbf{1}) \vee (-m\mathbf{1})$ a.e.;

$3°$ $|f_n - f| \wedge \mathbf{1} \longrightarrow 0$ a.e..

And if $\{f_n\} \subset C'_+$, the following are equivalent:

$1°$ $f_n \overset{u}{\longrightarrow} 0$ a.e.,

$2°$ $\mathbf{1}_{(f_n - \lambda \mathbf{1})^+} \longrightarrow 0$ a.e. for every $\lambda > 0$.

(9.3) (1) If $f_n \overset{u}{\longrightarrow} f$ a.e., then $\rho f_n \overset{u}{\longrightarrow} \rho f$ a.e. for every $\rho \in \mathbb{R}$.

(2) If $f_n \overset{u}{\longrightarrow} f$ a.e. and $g = f$ a.e., then $f_n \overset{u}{\longrightarrow} g$ a.e..

(3) If $f_n \overset{u}{\longrightarrow} f$ a.e. and $f_n \overset{u}{\longrightarrow} g$ a.e., then $g = f$ a.e..

(4) If $f_n \overset{u}{\longrightarrow} f$ a.e., $g_n \overset{u}{\longrightarrow} g$ a.e., and $g_n \le f_n$ a.e. for every n,
then $g \le f$ a.e..

(5) If $g_n \overset{u}{\longrightarrow} f$ a.e., $h_n \overset{u}{\longrightarrow} f$ a.e., and $g_n \le f_n \le h_n$ a.e. for
$n = 1, 2, ...,$ then $f_n \overset{u}{\longrightarrow} f$ a.e..

(6) If $f_n \overset{u}{\longrightarrow} f$ a.e. and $g_n \overset{u}{\longrightarrow} g$ a.e., then $f_n \vee g_n \overset{u}{\longrightarrow} f \vee g$ a.e.,
$f_n \wedge g_n \overset{u}{\longrightarrow} f \wedge g$ a.e., and $f_n + g_n \overset{u}{\longrightarrow} f + g$ a.e..

Contained in the above is the equivalence of the following:

$1°$ $f_n \overset{u}{\longrightarrow} f$ a.e.,

$2°$ $f_n - f \overset{u}{\longrightarrow} 0$ a.e..

Also

(9.4) The following are equivalent:

$1°$ $f_n \overset{u}{\longrightarrow} f$ a.e.;

$2°$ (i) $(f_n)^+ \xrightarrow{u} f^+$ a.e.,

(ii) $(f_n)^- \xrightarrow{u} f^-$ a.e..

And they imply

$$|f_n| \xrightarrow{u} |f| \text{ a.e..}$$

Moreover, if $f = 0$ a.e., the last statement is equivalent to $1°$ and $2°$.

(9.5)   If $C'' = G_1 \oplus G_2$, G_1 and G_2 bands, then for $\{f_n\} \subset C''$ and $f \in C''$, the following are equivalent:

$1°$ $f_n \xrightarrow{u} f$ a.e.;

$2°$ (i) $(f_n)_{G_1} \xrightarrow{u} f_{G_1}$ a.e.,

(ii) $(f_n)_{G_2} \xrightarrow{u} f_{G_2}$ a.e..

In particular, projection onto a band of C'' preserves almost everywhere unbounded order convergence of sequences.

§10. The Dieudonné topology

Each $\mu \in C'_+$ defines a seminorm $\| \cdot \|_\mu$ on C'' by $\|f\|_\mu = \langle |f|, \mu \rangle$ for every $f \in C''$. We denote by $|\sigma|(C'', J)$ the topology on C'' determined by the family $\{ \| \cdot \|_\mu \colon \mu \in J_+ \}$. Equivalently, $|\sigma|(C'', J)$ is the polar topology on C'' defined by the intervals of J. For a discussion of this topology, cf. [3; §22]. It has (among others) the following properties:

(1) It is locally convex;

(2) It is locally solid;

(3) Its null space is N;

(4)　J is the set of $|\sigma|(C'',J)$-continuous linear functionals on C'' (thus

$$\sigma(C'',J) \subset |\sigma|(C'',J) \subset \tau(C'',J),$$

where τ is the Mackey topology).

(5)　J is the set of linear functionals on C'' which are $|\sigma|(C'',J)$-continuous on the unit ball $[-\mathbf{1},\mathbf{1}]$ of C''.

(10.1)　If $f_n \longrightarrow f$ a.e., then $\{f_n\}$ $|\sigma|(C'',J)$-converges to f. The converse is false.

Proof.　We can assume $\{f_n\} \subset C''_+$ and $f=0$ (cf. (7.15)). Consider $\nu \in J_+$; we have to show that $\lim_n \langle f_n, \nu \rangle = 0$. By hypothesis, there exists a band H of C'' contained in N such that $(f_n)_{H^d} \longrightarrow 0$. Since ν is order continuous on C'', it follows $\lim_n \langle (f_n)_{H^d}, \nu \rangle = 0$. Now $\langle (f_n)_H, \nu \rangle = 0$ for every n, so

$$\lim_n \langle f_n, \nu \rangle = \lim_n \langle (f_n)_H, \nu \rangle + \lim_n \langle (f_n)_{H^d}, \nu \rangle = 0.$$

To show the converse is false, choose J such that $N \neq N$ (such J's exist) and set $f_n = 0$ $(n=1,2,...)$, $f=\mathbf{1}_N$. For every $\nu \in J$, $\langle f_n, \nu \rangle = 0$ for every n and $\langle f, \nu \rangle = 0$, so $\langle f, \nu \rangle = \lim_n \langle f_n, \nu \rangle$ trivially. However $\{f_n\}$ does not order converge to f a.e.: in effect, whatever band H of C'' we take in N, $f \notin H$, so $f_{H^d} \neq 0$; and since $(f_n)_{H^d} = 0$ for every n, $\{(f_n)_{H^d}\}$ does not order converge to f_{H^d}.

QED

We remark that the above implies that every $\nu \in J$ is continuous with respect to almost everywhere convergence of sequences.

In the remainder of this §, we assume $J = C'_\mu$, μ a fixed element of C'_+. As we stated earlier, we will in this case denote N, $\mathfrak{L}^\infty$, $\mathcal{N}$ by $N(\mu)$, $\mathfrak{L}^\infty(\mu)$, $\mathcal{N}(\mu)$ and write μ-a.e. for a.e.

We remarked in §6 that $N(\mu)$ can be described in terms of μ alone. In detail:

(10.2) For $f \in C''$, the following are equivalent:

    $1°$ $f \in N(\mu)$,

    $2°$ $\| f \|_\mu = 0$.

Proof. $1°$ of course implies $2°$. Assume $2°$ and consider $\nu \in (C'_\mu)_+$; we have to show $\langle f, \nu \rangle = 0$. assume first that $0 \leq \nu \leq n\mu$ for some n. then

$$0 \leq |\langle f, \nu \rangle| \leq \langle |f|, \nu \rangle \leq n\langle |f|, \mu \rangle = 0.$$

Now let ν be a general element of $(C'_\mu)_+$. then $\nu \wedge n\mu \uparrow \nu$ as $n \to \infty$. Since f is order continuous on C', $\langle f, \nu \rangle = \lim_n \langle f, \nu \wedge n\mu \rangle = 0$.

QED

Remark. The above is of course the statement that C'_μ is precisely the set of (Radon) measures absolutely continuous with respect to μ. The following is another form of this statement.

(10.3) C'_μ is the set of linear functionals on C'' which are $\|\cdot\|_\mu$-continuous on the unit ball $[-\mathbf{1}, \mathbf{1}]$ of C''.

This follows easily from [3; (27.5)].

One consequence:

(10.4)   For a sequence $\{f_n\}$ in C'' bounded μ-a.e. and $f \in C''$, the following are equivalent:

$1°$ $\{f_n\}$ $\|\cdot\|_\mu$-converges to f,

$2°$ $\{f_n\}$ $|\sigma|(C'', C'_\mu)$-converges to f.

§11. Convergence in Measure

(11.1)   Given $\mu \in C'_+$, the for every sequence $\{f_n\}$ in C'' bounded μ-a.e., $\lim_n \langle \mathbf{1}_{f_n}, \mu \rangle = 0$ implies $\lim_n \langle f_n, \mu \rangle = 0$. The converse is false.

Proof. By hypothesis, there exists a band H of C'' contained in $\mathcal{N}$ such that $\{(f_n)_{H^d}\}$ is bounded. Now $\mathbf{1}_{(f_n)_{H^d}} \le \mathbf{1}_{f_n}$ for every n, so $\lim_n \langle \mathbf{1}_{(f_n)_{H^d}}, \mu \rangle = 0$. Also $\langle (f_n)_{H^d}, \mu \rangle = \langle f_n, \mu \rangle$ for every n. Hence for simplicity, we can assume the sequence $\{f_n\}$ itself is bounded. Moreover, we can clearly assume $\{f_n\} \ge 0$.

Choose λ such that $\{f_n\} \le \lambda \mathbf{1}$. then $0 \le f_n \le \lambda \mathbf{1}_{f_n}$ for every n, so $\lim_n \langle f_n, \mu \rangle = 0$. To see that the converse is false, take $f_n = (1/n)\mathbf{1}$ $(n = 1, 2, ...)$ and $\mu \ne 0$.

QED

However, we have the following.

(11.2) Given $\mu \in C'_+$, then for a sequence $\{f_n\}$ in C''_+ bounded μ-a.e., the following are equivalent:

$\qquad$ 1° $\lim_n \langle f_n, \mu \rangle = 0$,

$\qquad$ 2° $\lim_n \langle f_n \wedge \mathbf{1}, \mu \rangle = 0$,

$\qquad$ 3° $\lim_n \langle \mathbf{1}_{(f_n - \lambda \mathbf{1})^+}, \mu \rangle = 0$ for every $\lambda > 0$.

Proof. Again we can assume $\{f_n\}$ is bounded. 1° implies 2°. Conversely, assume 2°; we show 1° holds. Choose $m \in \mathbb{N}$ such that $\{f_n\} \leq m\mathbf{1}$. Then for every n,

$$0 \leq \langle f_n, \mu \rangle = \langle f_n \wedge m\mathbf{1}, \mu \rangle \leq m\langle f_n \wedge \mathbf{1}, \mu \rangle,$$

hence $\lim_n \langle f_n, \mu \rangle = 0$.

Now assume 1°; we show 3° holds. Consider $\lambda \geq 0$. For each n,

$$0 \leq \mathbf{1}_{(f_n - \lambda \mathbf{1})^+} \leq (1/\lambda) f_n \qquad\qquad \text{(cf. [3; (17.9))},$$

hence

$$0 \leq \langle \mathbf{1}_{(f_n - \lambda \mathbf{1})^+}, \mu \rangle \leq (1/\lambda) \langle f_n, \mu \rangle,$$

hence

$$\lim_n \langle \mathbf{1}_{(f_n - \lambda \mathbf{1})^+}, \mu \rangle = 0.$$

Finally, assume 3°; we show 1° holds. Specifically, we show that for every $\lambda > 0$,

$$(*) \quad \limsup_n \langle f_n, \mu \rangle \leq \lambda \|\mu\|;$$

it will follow that $\limsup_n \langle f_n, \mu \rangle = 0$, that is, $1°$ holds. Fix $\lambda > 0$. By (11.1),

$$\lim_n \langle (f_n - \lambda\mathbf{1})^+, \mu \rangle = 0.$$

A fortiori,

$$\limsup_n \langle f_n - \lambda\mathbf{1}, \mu \rangle \leq 0.$$

Writing this in the form

$$\limsup_n \langle f_n, \mu \rangle - \lambda\langle \mathbf{1}, \mu \rangle \leq 0$$

gives us $(*)$.

$$\text{QED}$$

Note that the above can be written

(11.2a) Given $\mu \in C'_+$, then for a sequence $\{f_n\}$ in C''_+ bounded μ-a.e., the following are equivalent:

$1°$ $\lim_n \| f_n \|_\mu = 0,$

$2°$ $\lim_n \| f_n \wedge \mathbf{1} \|_\mu = 0,$

$3°$ $\lim_n \| \mathbf{1}_{(f_n - \lambda 1)^+} \|_\mu = 0$ for every $\lambda > 0$.

(11.2) in turn gives us

(11.3) Given $\mu \in C'_+$, then for a sequence $\{f_n\}$ in C''_+ (not necessarily bounded a.e.), the following are equivalent:

$1°$ $\lim_n \langle f_n \wedge \mathbf{1}, \mu \rangle = 0,$

$2°$ $\lim_n \langle \mathbf{1}_{(f_n - \lambda 1)^+}, \mu \rangle = 0$ for every $\lambda > 0$.

This can be derived from (11.2) using the following. If $\lambda < \kappa$, then

(1) $f \wedge \lambda \mathbf{1} = (f \wedge \kappa \mathbf{1}) \wedge \lambda \mathbf{1}$,

(2) $\mathbf{1}_{(f - \lambda 1)^+} = \mathbf{1}_{(f \wedge \kappa 1 - \lambda 1)^+}.$

If one, hence also the other, of the conditions in (11.3) holds, we will write

$$f_n \longrightarrow 0 \ in \ \mu\text{-}Measure.$$

More generally, for a sequence $\{f_n\}$ in C'' and $f \in C''$, we write write

$$f_n \longrightarrow f \ in \ \mu\text{-}Measure$$

if $|f_n - f| \longrightarrow 0$ in μ-Measure.

Note that by (11.2a), for sequences bounded μ-a.e., convergence in μ-Measure reduces to convergence in the seminorm $\| \cdot \|_\mu$. We state this formally:

(11.4) Given $\mu \in C'_+$, then for a sequence $\{f_n\}$ in C'' bounded μ-a.e., the following are equivalent:

   $1°$ $f_n \longrightarrow f$ in μ-Measure,

   $2°$ $\lim_n \| f_n - f \|_\mu = 0$.

We extend our definition to the case where our band J is not necessarily generated by a single μ.

Given a sequence $\{f_n\}$ in C'' and $f \in C''$, we write write $f_n \longrightarrow f$ in J-*Measure*, or simply *in Measure*, if $f_n \longrightarrow f$ in μ-measure for every

$\mu \in J_+$. For sequences bounded a.e., convergence in J-Measure reduces to topological convergence:

(11.5) For a sequence $\{f_n\}$ in C'' bounded a.e., the following are equivalent:

      $1°$ $f_n \longrightarrow f$ in J-Measure,

      $2°$ $\{f_n\}$ $|\sigma|(C'', J)$ — converges to f.

This is an immediate consequence of (11.4).

Remarks. (1) Note the upper-case M in convergence in Measure. This is to distinguish it from the convergence in measure in standard integration theory in $\ell^\infty(X)$. We will see later that our above convergence gives the latter under the projection of C'' onto C''_a.

(2) Note also the resemblance between convergence in Measure and unbounded order convergence (cf. (8.6) and the comment following (8.2)).

(10.4) gives us

(11.6) Given $\mu \in C'_+$, then for a sequence $\{f_n\}$ in C'' and $f \in C''$, the following are equivalent:

      $1°$ $f_n \longrightarrow f$ in C'_μ-Measure,

      $2°$ $f_n \longrightarrow f$ in μ-Measure.

It follows easily from this that any proposition on convergence in Measure (that is, in J-Measure) holds also for convergence in μ-Measure, and conversely.

Almost everywhere unbounded order convergence is strictly stronger than convergence in Measure:

(11.7) If $f_n \xrightarrow{u} f$ a.e., then $f_n \longrightarrow f$ in Measure. The converse is false.

This follows easily from (10.1). We leave the details to the reader.

(11.8) (1) If $f_n \longrightarrow f$ in Measure, then $\rho f_n \longrightarrow \rho f$ in Measure for every $\rho \in \mathbb{R}$.

(2)   If $f^n \longrightarrow f$ in Measure and $g - f \in N$, then $f_n \longrightarrow g$ in Measure.

(3)   If $f_n \longrightarrow f$ in Measure and $f_n \longrightarrow g$ in Measure, then $g - f \in N$.

(4)   If $f_n \longrightarrow f$ in measure and $g_n - f_n \in N$ $(n = 1, 2, ...)$, then $g_n \longrightarrow f$ in Measure.

(5)   If $f_n \longrightarrow f$ in Measure and $g_n \longrightarrow g$ in Measure, then $f_n + g_n \longrightarrow f + g$ in Measure, $f_n \vee g_n \longrightarrow f \vee g$ in Measure, and $f_n \wedge g_n \longrightarrow f - g$ in Measure.

We leave the verifications to the reader.

(11.9) *Corollary.* The following are equivalent:

$\quad$ 1° $f_n \longrightarrow f$ in Measure;

$\quad$ 2° (i) $(f_n)^+ \longrightarrow f^+$ in Measure,

$\quad\quad$ (ii) $(f_n)^- \longrightarrow f^-$ in Measure.

And they imply

$$|f_n| \longrightarrow |f| \text{ in Measure.}$$

Moreover, if $f = 0$, the last statement is equivalent to 1° and 2°,

(1.10) If $C'' = G_1 \oplus G_2$, G_1 and G_2 bands, then for $\{f_n\} \subset C''$ and $f \in C''$, the following are equivalent:

$\quad$ 1° $f_n \longrightarrow f$ in Measure;

$\quad$ 2° (i) $(f_n)_{G_1} \longrightarrow f_{G_1}$ in Measure,

$\quad\quad$ (ii) $(f_n)_{G_2} \longrightarrow f_{G_2}$ in Measure.

§12. Sequences in $\mathfrak{L}^\infty$

The following strengthens (4.1) and (5.1).

(12.1) $\mathfrak{L}^\infty$ and $\mathcal{N}$ are closed in C'' under almost everywhere unbounded order convergence of sequences.

Proof. Consider $\{f_n\} \subset \mathfrak{L}^\infty$, and suppose first that $f_n \longrightarrow f$ a.e.. So there exists a band H of C'' contained in $\mathcal{N}$ such that

$\quad$ (i) $(f_n)_{H^d} \longrightarrow f_{H^d}$.

For every n, $(f_n)_H \in H \subset \mathcal{N} \subset \mathfrak{L}^\infty$, so

$$(f_n)_{H^d} = f_n - (f_n)_H \in \mathfrak{L}^\infty.$$

It follows from (4.1) and (i) above that $f_{H^d} \in \mathfrak{L}$, hence $f = f_H + f_{H^d} \in \mathfrak{L}^\infty$.

Now suppose only that $f_n \xrightarrow{u} f$ a.e.. So by (9.1), we have in particular that for each m=1, 2, ...,

$$(f_n \wedge m\mathbf{1}) \vee (-m\mathbf{1}) \longrightarrow (f \wedge m\mathbf{1}) \vee (-m\mathbf{1}) \text{ a.e.,}$$

hence by the first part of the proof, $(f \wedge m\mathbf{1}) \vee (-m\mathbf{1}) \in \mathfrak{L}^\infty$. Since $(f \wedge m\mathbf{1}) \vee (-m\mathbf{1}) = f$ for m sufficiently large, $f \in \mathfrak{L}^\infty$.

We turn to $\mathcal{N}$. Suppose $\{f_n\} \subset \mathcal{N}$ and $f_n \xrightarrow{u} f$ a.e.; we show $f \in \mathcal{N}$. By the preceding part of the proof, $f \in \mathfrak{L}^\infty$. We show $f \in N$, hence $f \in \mathfrak{L}^\infty \cap N = \mathcal{N}$. By (9.5), $f_n \xrightarrow{u} f_N$ a.e., so by (9.3), $f = f_N$ a.e.. This is equivalent to $f - f_N \in \mathcal{N}$. Thus $f = f_N + (f - f_N) \in N$.

QED

Remark. $\mathfrak{L}^\infty$ and $\mathcal{N}$ are, in general, not closed in C'' under $|\sigma|(C'', J)$–convergence, even of sequences, since the topology $|\sigma|(C'', J)$ cannot distinguish between $\mathcal{N}$ and N. A fortiori, they are, in general, not closed under convergence in Measure of sequences.

In the above proof, we used the fact (cf. (9.5) and the comment following it) that $f_n \xrightarrow{u} f$ a.e. implies $(f_n)_G \xrightarrow{u} f_G$ a.e. for every band G of C''. (1) below says that for the band I dual to J, we have a stronger conclusion.

(12.2) Given $\{f_n\} \subset C''$ and $f \in \mathcal{C}''$,

    (1) If $f_n \xrightarrow{u} f$ a.e., then $(f_n)_I \xrightarrow{u} f_I$.

    (2) If $\{f_n\} \subset \mathfrak{L}^\infty$, then also the converse holds.

Proof. (1) Assume $f_n \xrightarrow{u} f$ a.e.. So there exists a band H of C'' contained in $\mathcal{N}$ such that $(f_n)_{H^d} \xrightarrow{u} f_{H^d}$. Since $I \subset H^d$, (8.5) gives the desired result.

    (2)

Lemma. For $g, h \in \mathfrak{L}^\infty$, the following are equivalent:

    $1°$ $g \le h$ a.e.,

    $2°$ $g_I \le h_I$.

A fortiori, this holds with " $\le$ " replaced by "$=$".

That $1°$ implies $2°$ follows by the same argument as was used for (1) above (that is, $I \subset H^d$). Assume $2°$. So $(g_I - h_I)^+ = 0$. Writing this $[(g-h)^+]_I = 0$, we have $(g-f)^+ \in N$, hence (since $(g-f)^+ \in \mathfrak{L}^\infty$) $(g-f)^+ \in \mathcal{N}$. It follows $g \le h$ a.e. (cf. the Scholium at the end of §7). This establishes the Lemma; we proceed with the proof of (2).

For simplicity (cf. (8.2), (8.3), and (9.2), (9.3)), we can take $\{f_n\} \subset (\mathfrak{L}^\infty)_+$ and $f = 0$. Thus (2) can be written $(f_n)_I \wedge \mathbf{1} \longrightarrow 0$, and we have to show $f_n \wedge \mathbf{1} \longrightarrow 0$ a.e.. Since $(f_n)_I \wedge \mathbf{1} = (f_n \wedge \mathbf{1})_I$, this reduces to showing:

> $(*)$ For a bounded sequence $\{k_n\} \subset (\mathfrak{L}^\infty)_+$,
>
> $(k_n)_I \longrightarrow 0$ implies $k_n \longrightarrow 0$ a.e..

Set $k = \limsup_n (k_n)$. Then

> (i) $k \in \mathfrak{L}^\infty$,
>
> (ii) $k_I = \limsup_n (k_n)_I$.

By (ii) and the hypothesis of $(*)$, $k_I = 0$; hence by (i) and the Lemma, $k = 0$ a.e.. Thus there exists a band H of C'' contained in $\mathcal{N}$ such that $k_{H^d} = 0$. Thus

$$\limsup_n (k_n)_{H^d} = k_{H^d} = 0,$$

whence $\lim_n (k_n)_{H^d} = 0$, and we have $\lim_n k_n = 0$ a.e..

$$\text{QED}$$

Chapter 3. Some Classical Theorems

In this paper, by "standard Lebesgue theory", we will mean the theory of Lebesgue integration of bounded functions on X with respect to some positive Radon (equivalently, regular) measure μ. Recall that we have identified the space $\ell^\infty(X)$ of bounded functions on X with the band C''_a of C''.

We will see in the next chapter that for each $\mu \in C'_+$, the image $(\mathfrak{L}^\infty(\mu))_a$ of $\mathfrak{L}^\infty(\mu)$ under the projection of C'' onto C''_a is precisely the set of (bounded) functions on X μ-integrable in the standard sense. (And for our our general band J, $(\mathfrak{L}^\infty)_a$ is the set of bounded functions on X ν-integrable in the standard sense for every $\nu \in J_+$.)

In the present chapter, we establish the corresponds for $\mathfrak{L}^\infty$ to some classical theorems of standard Lebesgue theory. We show in the next chapter that under the projection of C'' onto C''_a, our theorems give the classical ones. As indicated in the Introduction, the role played by $\int f d\mu$ in $\ell^\infty(X)$ is played in C'' by the bilinear form $\langle f, \mu \rangle$.

In a succeeding paper, we carry out the same program for the standard theory of measurable functions (not necessarily bounded) on X. This requires first identifying a superspace of C'' which plays the

62

same role with respect to C'' that $\mathbb{R}^X$ plays with respect to $\ell^\infty(X)$.

As in the preceding part of the paper, J is a fixed band of C', and $\mathfrak{L}^\infty = \mathfrak{L}^\infty(J)$. We will designate our theorems by the names of the theorems in standard Lebesgue theory to which they correspond.

§13. Elementary theorems

The following corresponds to the elementary fact that a bounded function on X is μ-integrable (in the standard sense) if and only if it is μ-measurable.

(13.1) For $f \in C''$, the following are equivalent:

$\quad 1^\circ\ f \in \mathfrak{L}^\infty$,

$\quad 2^\circ\ \mathbf{1}_{(f-\lambda 1)^+} \in \mathfrak{L}^\infty$ for every $\lambda \in \mathbb{R}$.

This is actually a property of every σ-order closed Riesz subspace of C'' containing $\mathbf{1}$. It follows from the Freudenthal theorem [3; (17.10)].

Given a Riesz space E, we denote by E^b the Riesz space of all order bounded linear functionals on E. The essence of Fatou's Lemma is a property of E^b for every Riesz space E. We state it on C'' (that is, $E = C'$), and to emphasize that in this general situation, countability does not play a role, we state it for a net.

(13.2)   (Fatou)   If $\{f_\alpha\}$ is a bounded net in C'', then for every $\mu \in C'_+$,

$$\langle \liminf_\alpha(f_\alpha), \mu \rangle \;\leq\; \liminf_\alpha \langle f_\alpha, \mu \rangle$$
$$\leq\; \limsup_\alpha \langle f_\alpha, \mu \rangle$$
$$\leq\; \langle \limsup_\alpha(f_\alpha), \mu \rangle.$$

Proof.   We establish the first inequality.   Set $g = \liminf_\alpha(f_\alpha)$; we have to show

(i) $\langle g, \mu \rangle \leq \liminf_\alpha \langle f_\alpha, \mu \rangle$.

For each α, set $g_\alpha = \bigwedge_{\beta \geq \alpha} f_\beta$.   Then:

(ii) for each α, $\langle g^\alpha, \mu \rangle \leq \inf_{\beta \geq \alpha} \langle f_\beta, \mu \rangle$;

(iii) $\langle g, \mu \rangle = \sup_\alpha \langle g_\alpha, \mu \rangle$.

(iv)   For every $\beta \geq \alpha$, $g_\alpha \leq f_\beta$, so $\langle g_\alpha, \mu \rangle \leq \langle f_\beta, \mu \rangle$; hence $\langle g_\alpha, \mu \rangle \leq \inf_{\beta \geq \alpha} \langle f_\beta, \mu \rangle$.

(v)   This follows from $g_\alpha \uparrow g$ and the fact that μ is order continuous on C''.

We thus have

$$\langle g, \mu \rangle = \sup_\alpha \langle g_\alpha, \mu \rangle \qquad\qquad\qquad \text{(iii)}$$
$$\leq \sup_\alpha \left(\inf_{\beta \geq \alpha} \langle f_\beta, \mu \rangle \right) \qquad\qquad \text{(ii)}$$
$$= \liminf_\alpha \langle f_\alpha, \mu \rangle.$$

$$\text{QED}$$

A classical theorem of F. Riesz states that if a sequence of Lebesgue integrable functions converges in measure, then some subsequence order converges almost everywhere to the same limit. We give the correspond to this for $\mathfrak{L}^{\infty}$.

(13.3)   (Riesz)   Given $\mu \in C'_+$, $\{f_n\} \subset \mathfrak{L}^{\infty}(\mu)$, and $f \in \mathfrak{L}^{\infty}(\mu)$.   If $f_n \longrightarrow f$ in μ-Measure, then some subsequence of $\{f_n\}$ order converges unboundedly to f μ-a.e..

Proof.

Lemma 1. If a bounded sequence $\{g_n\}$ in C''_μ converges to $g \in C''_\mu$ in the norm $\| \cdot \|_\mu$, then some subsequence order converges to g.

We can assume $\{g_n\} \geq 0$ and $g = 0$. By taking a subsequence if necessary, we can assume $\| g_n \|_\mu \leq 1/2^n$ (n=1,2,...).   Set $g = \limsup_n g_n$; we show $g = 0$. For each n set $h_n = \bigvee_{m \geq n} g_m$. Then

$$\text{(i)}   \| h_n \|_\mu \leq 1/2^n,$$
$$\text{(ii)} g = \bigwedge_n h_n.$$

It follows $\| g \|_\mu = 0$, hence, since $\| \cdot \|_\mu$ is a norm on C''_μ, $g = 0$.

Lemma 2. If a sequence $\{k_n\}$ in C''_μ converges in μ-Measure to $k \in C''_\mu$, then some subsequence order converges unboundedly to k.

We can again assume $\{k_n\} \geq 0$ and $k = 0$. So the hypothesis states that $\lim_n \| k_n \wedge \mathbf{1} \|_\mu = 0$ (cf. (11.2a)). It follows from Lemma 1 that there exists a subsequence $\{k_{n_m} \wedge \mathbf{1}\}$ such that $k_{n_m} \wedge \mathbf{1} \to 0$ (as $m \to \infty$). Thus $k_{n_m} \xrightarrow{u} 0$.

We proceed to prove the theorem. By (11.10), $(f_n)_\mu \to f_\mu$ in μ-Measure. Hence by Lemma 2, there exists a subsequence $\{(f_{n_m})_\mu : m = 1, 2, ...\}$ such that $(f_{n_m})_\mu \xrightarrow{u} f_\mu$ (as $m \to \infty$). It follows from (12.2) that $f_{n_m} \xrightarrow{u} f$ μ-a.e. (as $m \to \infty$).

$$\text{QED}$$

We next give the correspond in $\mathfrak{L}^\infty$ of the Lebesgue Dominated Convergence Theorem. (Of course, in $\mathfrak{L}^\infty$, "dominated" is equivalent to "bounded".) Indeed, it holds for $\mathfrak{L}^\infty(J)$, not only for $\mathfrak{L}^\infty(\mu)$, and we state it in this more general form. Actually, we already have this theorem, as we show.

(13.4) (Lebesgue) Let $\{f_n\}$ be a sequence in $\mathfrak{L}^\infty$ bounded a.e.. If $f_n \to f$ a.e., then:

(1) $f \in \mathfrak{L}^\infty$,;

(2) for every $\nu \in J_+$, $\lim_n \langle | f_n - f |, \nu \rangle = 0$ (hence $\langle f, \nu \rangle = \lim_n \langle f_n, \nu \rangle$).

(1) is contained in (12.1), and (2) follows from the comment following (10.1)

§14. The Egorov Theorem

The theorems in this § correspond to the Egorov theorem for bounded integrable functions. In essence, they are about elements of C'', not merely those of $\mathfrak{L}^\infty$. To emphasize this, we first establish them for a general net in C'' ((14.1) – (14.3)). In the following, it will be understood without mention that any earlier result referred to can be shown to hold for uncountable sets (resp. for nets) even though stated only for countable sets (resp. sequences).

(14.1) Let μ be a fixed element of C'_+. Suppose $\{f_\alpha\}$ is a net in C'' such that

$$f_\alpha \xrightarrow{\text{u}} f \quad \mu\text{-a.e.}.$$

Then for each $\lambda > 0$ and $\delta > 0$ in $\mathbb{R}$, there exist components of $\mathbf{1}$, $d = d(\lambda, \delta)$ and $e = e(\lambda, \delta)$, and an index $\alpha_0 = \alpha(\lambda, \delta)$ such that:

(1) $e + d = \mathbf{1}$ (so $e \wedge d = 0$);

(2) $\langle d, \mu \rangle \leq \delta$;

(3) $|(f_\alpha)_e - f_e| \leq \lambda \mathbf{1}$ for every $\alpha \geq \alpha_0$.

Moreover, if G is a band of C'' containing $\{f_\alpha\}$ and f, then the conclusion holds with $\mathbf{1}$ replaced by $\mathbf{1}_G$.

Proof. There is no loss of generality in assuming $\{f_\alpha\} \geq 0$ and $f = 0$ (cf. (9.4) and the comment preceding it). We first prove the proposition for $f_\alpha \xrightarrow{\text{u}} f$ (not merely μ-a.e.).

Case I, $f_\alpha \downarrow 0$: For every α, $\mathbf{1}_{(f_\alpha - \lambda \mathbf{1})^+} \leq (1/\lambda) f_\alpha$. Moreover, since $\{f_\alpha\}$ is a descending net, $\{\mathbf{1}_{(f_\alpha-\lambda\mathbf{1})+}\}$ is one also. Thus

$$\mathbf{1}_{(f_\alpha-\lambda\mathbf{1})+}\downarrow 0.$$

This gives us in turn that the net of real numbers $\{\langle\mathbf{1}_{(f_\alpha-\lambda\mathbf{1})+},\mu\rangle\}$ is also descending with limit 0. Choose α_0 such that

$$\langle\mathbf{1}_{(f_{\alpha_0}-\lambda\mathbf{1})+},\mu\rangle \le \delta,$$

and set

$$d = \mathbf{1}_{(f_{\alpha_0}-\lambda\mathbf{1})+}, \quad e=\mathbf{1}-d.$$

d and e satisfy (1) and (2), and for every $\alpha \ge \alpha_0$, $0\le f_\alpha\le f_{\alpha_0}$, so $0 \le (f_\alpha)_e \le (f_{\alpha_0})_e \le \lambda\mathbf{1}$. Thus (3) is also satisfied.

Case II, $\{f_\alpha\}$ bounded: The hypothesis reduces to $f_\alpha\rightarrow 0$, so there exists a net $\{g_\alpha\}$ such that $g_\alpha\downarrow 0$ and $0\le f_\alpha\le g_\alpha$ for every α. By Case I, there exist d, e, and α_0 satisfying (1)—(3) for $\{g_\alpha\}$. Then they do also for $\{f_\alpha\}$.

Case III, $\{f_\alpha\}$ not necessarily bounded: Choose $\kappa>\lambda$. $f_\alpha\wedge\kappa\mathbf{1}\rightarrow 0$, so by Case II, there exist d, e, and α_0 satisfying (1)—(3) for $\{f_\alpha\wedge\kappa\mathbf{1}\}$. In particular, for $\alpha\ge\alpha_0$, $(f_\alpha\wedge\kappa\mathbf{1})_e \le \lambda\mathbf{1}$. This can be written $(f_\alpha)_e\wedge\kappa\mathbf{1}\le\lambda\mathbf{1}$, hence the desired inequality, $(f_\alpha)_e\le\lambda\mathbf{1}$, follows from the

Lemma. If $\lambda<\kappa$, then for every $f\in C''$, $f\wedge\kappa\mathbf{1}\le\lambda\mathbf{1}$ implies $f\le\lambda\mathbf{1}$.

We use the easily verified identity:

$$(f - \lambda \mathbf{1})^+ \wedge (\kappa - \lambda)\mathbf{1} = (f \wedge \kappa\mathbf{1} - \lambda\mathbf{1})^+.$$

By the hypothesis of the Lemma, the right side is 0, so the left side is also. Since $(\kappa - \lambda)\mathbf{1}$ is an order unit for C'', it follows $(f - \lambda\mathbf{1})^+ = 0$.

The final statement of the theorem is easily verified (for $f_\alpha \overset{u}{\to} 0$) by replacing d by $d \wedge \mathbf{1}_G$ and e by $e \wedge \mathbf{1}_G$. This establishes the theorem for $f_\alpha \overset{u}{\to} 0$.

Now assume $f_\alpha \overset{u}{\to} 0$ μ-a.e.. So there exists a band H of C'' contained in $\mathcal{N}(\mu)$ such that $(f_\alpha)_{H^d} \overset{u}{\to} 0$. Then by the above, there exist d, e, and α_0 satisfying (1) — (3) for $\{(f_\alpha)_{H^d}\}$ with $\mathbf{1}$ replaced by $\mathbf{1}_{H^d}$. Set $d' = d + \mathbf{1}_H$. Then d', e, and α_0 satisfy (1) — (3) for $\{f_\alpha\}$. Again, the final statement is easily verified.

QED

(14.2) (First Egorov Theorem for C'') Given $\mu \in C'_+$, suppose $\{f_\alpha\}$ is a net in C'' such that

$$f_\alpha \overset{u}{\to} f \quad \mu\text{-a.e.}.$$

Then for each $\delta > 0$, there exist components of $\mathbf{1}$, $d = d(\delta)$ and $e = e(\delta)$ such that

(1) $d + e = \mathbf{1}$,

(2) $\langle d, \mu \rangle \leq \delta$,

(3) $\lim_\alpha \| (f_\alpha)_e - f_e \| = 0$.

And if G is a band of C'' containing $\{f_\alpha\}$ and f, the conclusion holds with $\mathbf{1}$ replaced by $\mathbf{1}_G$.

Proof. For each $p = 1, 2, ...$, set $\lambda_p = 1/p$, $\delta_p = \delta/2^p$, and let d_p, e_p be given by (14.1). Then $d = \vee_p d_p$ and $e = \wedge_p e_p$ have the desired properties.

QED

(14.3) (Second Egorov Theorem for C'') Given $\mu \in C'_+$, suppose $\{f_\alpha\}$ is a net in C'' such that

$$f_\alpha \xrightarrow{u} f \ \mu\text{-a.e.}.$$

Then there exist components of $\mathbf{1}$ $\{e_m \colon m=0, 1, 2, ...\}$ such that

(1) $e_{m_1} \wedge e_{m_2} = 0$ for $m_1 \neq m_2$,

(2) $\vee_m e_m = \mathbf{1}$,

(3) $\langle e_0, \mu \rangle = 0$,

(4) $\lim_\alpha \| (f_\alpha)_{e_m} - f_{e_m} \| = 0$ for $m = 1, 2, ...$.

Proof. Let $\mathbf{1} = d_1 + e_1$ be the decomposition of $\mathbf{1}$ obtained in (14.2) for $\delta = 1$. $(f_\alpha)_{d_1} \xrightarrow{u} f_{d_1}$ μ-a.e. (cf. (9.5)), so we can apply (14.2) again to obtain a decomposition $d_1 = d_2 + e_2$ for $\delta = 1/2$. Continuing in this fashion, and setting $e_0 = \wedge_m d_m$ gives us the desired result.

QED

The Egorov theorems for $\mathfrak{L}^\infty(\mu)$ (instead of C'') hold only for sequences:

(14.4) (First Egorov Theorem for $\mathfrak{L}^\infty(\mu)$)   Given $\mu \in C'_+$ and $\{f_n\} \subset \mathfrak{L}^\infty(\mu)$, suppose $f_n \xrightarrow{u} f$ μ-a.e. (so $f \in \mathfrak{L}^\infty(\mu)$ (12.1)). Then for each $\delta > 0$, there exist components d and e of $\mathbf{1}$ such that:

> (1) $d, e \in \mathfrak{L}^\infty(\mu)$,
>
> (2) $d + e = \mathbf{1}$,
>
> (3) $\langle d, \mu \rangle \leq \delta$,
>
> (4) $\lim_n \| (f_n)_e - f_e \| = 0$.

Moreover, if G is a band of C'' containing $\{f_n\}$ and such that $\mathbf{1}_G \in \mathfrak{L}^\infty(\mu)$, then the conclusion holds with $\mathbf{1}$ replaced by $\mathbf{1}_G$.

The proof is the same as for (14.2) (that $d \in \mathfrak{L}^\infty(\mu)$ follows from the σ-order closedness of $\mathfrak{L}^\infty(\mu)$).

(14.4) (Second Egorov Theorem for $\mathfrak{L}^\infty(\mu)$)   Given $\mu \in C'_+$ and $\{f_n\} \subset \mathfrak{L}^\infty(\mu)$, suppose $f_n \longrightarrow f$ μ-a.e.. Then there exist components of $\mathbf{1}$ $\{e_m \colon m = 0, 1, 2, ...\}$ such that:

> (1) $e_m \in \mathfrak{L}^\infty(\mu)$ for every m,
>
> (2) $e_{m_1} \wedge e_{m_2} = 0$ for $m_1 \neq m_2$,
>
> (3) $\vee_m e_m = \mathbf{1}$,
>
> (4) $\langle e_o, \mu \rangle = 0$,
>
> (5) $\lim_n \| (f_n)_{e_m} - f_{e_m} \| = 0$ for $m = 1, 2, ...$.

The proof is the same as for (14.3).

§15. The Lusin Theorem

Lusin's Theorem reads as follows: Given a regular measure μ on X and a μ-measurable function f on X, then for each $\delta > 0$, f is continuous on some closed set (depending on f and δ) whose complement has μ-measure $\leq \delta$.

(15.1) (The Lusin Theorem for $\mathfrak{L}^{\infty}(\mu)$)   Given $\mu \in C'_+$, then for $f \in C''$, the following are equivalent:

   $1°$ $f \in \mathfrak{L}^{\infty}(\mu)$;

   $2°$ there exists a bounded sequence $\{g_n\}$ in C such that $g_n \longrightarrow f$ μ-a.e.;

   $3°$ for each $\delta > 0$, there exists a usc component e of $\mathbf{1}$ such that:

   (1) $\langle \mathbf{1} - e, \mu \rangle \leq \delta$,

   (2) $f_e \in C_e$;

   $4°$ there exist components $\{e_n\}$ of $\mathbf{1}$ such that:

   (1) e_n is a usc element for every n,

   (2) $e_n \uparrow \mathbf{1}$ μ-a.e.,

   (3) $f_{e_n} \in C_{e_n}$ for every n.

Proof. For simplicity, we can take $0 \leq f \leq \mathbf{1}$. Assume $1°$. By [3; (41.2)], there exists a bounded sequence $\{g_n\} \subset C$ such that $(g_n)_\mu \longrightarrow f_\mu$. It follows from (10.2) that $2°$ holds.

Assume 2°.   Consider $\delta > 0$ and let $\mathbf{1} = d + e$ be the decomposition of $\mathbf{1}$ given by (14.4). Since $\{g_n\} \subset C$, d is an lsc element (cf. the proof of (14.1)), hence e is a usc element. We show $f_e \in C_e$ (hence 3° holds). $\{(g_n)_e\} \subset C_e$, which is norm closed [3; (18.3)]. Since $\lim_n \| (g_n)_e - f_e \| = 0$, it follows $f_e \in C_e$.

Assume 3°. Setting $\delta_n = 1/n$ $(n = 1, 2, ...)$, we obtain a sequence $\{e_n\}$ of usc components of $\mathbf{1}$ such that for each n,

(i) $\langle \mathbf{1} - e_n, \mu \rangle \leq 1/n$,

(ii) $f_{e_n} \in C_{e_n}$.

As is easily verified, we can arrange to have $e_1 \leq e_2 \leq ....$ Set $e = \bigvee_n e_n$ and $d = \mathbf{1} - e$. Then $\langle d, \mu \rangle = 0$, so $e = \mathbf{1}$ μ-a.e.. Thus 4° holds.

Finally, assume 4°. We show first that $f_{e_n} \in \mathfrak{L}^\infty(\mu)$ for every n. Fix n. By (3), $f_{e_n} = g_{e_n}$ for some $g \in C$, and since $0 \leq f \leq \mathbf{1}$, we can choose $0 \leq g \leq \mathbf{1}$.   Then $g_{e_n} = g \wedge e_n \in \mathfrak{L}^\infty(\mu)$.   Thus $f_{e_n} = g_{e_n} \in \mathfrak{L}^\infty(\mu)$.   Now $e_n \uparrow \mathbf{1}$ μ-a.e., so by (7.13), $f_{e_n} = f \wedge e_n \uparrow f \wedge \mathbf{1} = f \, \mu$-a.e.. It follows from (12.1) that $f \in \mathfrak{L}^\infty(\mu)$, and we have 1°.

QED

Chapter 4. The Projection of C'' onto C''_a

Our final goal is to show that under the projection of C'' onto C''_a $(= \ell^\infty(X))$, $\mathfrak{L}^\infty(\mu)$ and its properties give us the space of bounded functions on X which are μ-integrable in the standard sense, together with their properties. We do this is Chapter 5.

In the present chapter, we examine the projection itself. Under the projection, the inverse image of each $f \in C''_a$ is the translate of C''_d determined by f. Consequently, our attention will largely be centered on these translates.

By a *translate* of C''_d, we mean a set of the form $C''_d + f$, f an element of C''. We will say the translate is *determined by* f. We will use without comment the following (trivial) properties of translates.

(1) Every $f \in C''$ lies in the translate which it determines.

(2) Conversely, a translate is determined by each element in it.

(3) Two elements of C'' lie in the same translate if and only if they differ by an element of C''_d.

(4) C''_d is itself a translate, the one determined by 0.

74

(5)   The translates are mutually disjoint (as sets) with union C''.

(6)   Each translate contains exactly one element of C''_a.

(7)   If f and g lie in the same translate and $g \leq f$, then the interval $[g, f]$ is contained in the translate.

(8)   Every translate is order closed.

It follows from (6) and (2) that every translate can be written in the form $C''_d + f$, f the unique element of C''_a lying in it. From now on, unless otherwise stated, we will write the translates in this form, that is, when we speak of a translate $C''_d + f$, it will be understood that $f \in C''_a$.

The projection of C'' onto C''_a is defined by $g \mapsto g_a$, g running through C''. So for $f \in C''_a$, the inverse image of under the projection is the set of all g such that $g_a = f$. This set is precisely the translate $C''_d + f$. It follows from (2.14) that for every $f \in C''_a$, $u(f)$ and $\ell(f)$ also lie in the translate $C''_d + f$.

§16. The star elements

(16.1) Given a translate A of C''_d, then for every ℓ-element g and u-element h in A, $g \leq h$.

Proof. By definition, there exist subsets $\{g_\alpha\}$, $\{h_\beta\}$ in U such that $g = \vee_\alpha g_\alpha$, $h = \wedge_\beta h_\beta$. We show that $g_\alpha \le h_\beta$ for every α and β; it will follow that $g \le h$. $g_\alpha \le g$ and $h_\beta \ge h$, so

$$(g_\alpha)_a \le g_a = h_a \le (h_\beta)_a.$$

Since the projection of U onto U_a is a (Riesz space) isomorphism [3; (53.1)], $g_\alpha \le h_\beta$.

$$\text{QED}$$

Now consider $f \in C_a''$. The translate $C_d'' + f$ contains at least one ℓ-element and one u-element (for example, $\ell(f)$ and $u(f)$). Hence by (16.1) and (2.1), it contains a largest ℓ-element and a smallest u-element, with the former $\le$ the latter. We denote them by f_* and f^* respectively. Summing up:

For each $f \in C_a''$:

    (I) $f_* = $ the largest ℓ-element in $C_d'' + f$.

    (II) $f^* = $ the smallest u-element in $C_d'' + f$.

    (III) $f_* \le f^*$.

It is clear that $u(k) = f^*$ for every $k \in C_d'' + f$, $k \le f^*$ and $\ell(k) = f_*$ for every $k \in C_d'' + f$, $k \ge f_*$. Thus for $f_* \le k \le f^*$, $u(k) = f^*$ and $\ell(k) = f_*$. In particular, $u(f_*) = f^*$ and $\ell(f^*) = f_*$. Finally, from the isomorphism of U with U_a, $f_* = f^*$ (hence) $\in$ U if and only if $f \in U_a$. (In particular, $(\mathbf{1}_a)_* = (\mathbf{1}_a)^* = \mathbf{1}$ and $O_* = O^* = O$. We will use these identities without explicit mention.)

Two immediate characterizations of f_* and f^*:

(16.2) Given $f \in C_a''$:

    (1) $f^* = \wedge \{h \in U: h_a \geq f\}$,

       $f_* = \vee \{g \in U: g_a \leq f\}$.

    (2) $f^* = \wedge \{h$ an lsc element: $h_a \geq f\}$,

       $f_* = \vee \{g$ a usc element: $g_a \leq f\}$.

We leave the verification to the reader.

We record the properties of the star elements. The reader will note the similarity of many of them with properties of $\ell(f)$ and $u(f)$ (the latter are of course defined on all of C'', not just on C_a''). Consequently, we will omit not only those proofs which are straightforward, but also any proof which is similar to one for $\ell(f)$ or $u(f)$.

(16.3) (1) For $f, g \in C_a''$, the following are equivalent:

      $1°$ $g \leq f$,

      $2°$ $g^* \leq f^*$,

      $3°$ $g_* \leq f_*$.

    (2) For every $f \in C_a''$,

$$-f^* = (-f)_*.$$

    (3) For every $f \in C_a''$ and $\rho \in \mathbb{R}_+$,

$$(\rho f)^* = \rho f^*,$$

$$(\rho f)_* = \rho f_*.$$

(1) above gives us

(16.4) For $f \in C_a''$, the following are equivalent:

$$1° \quad \lambda \mathbf{1}_a \leq f \leq k \mathbf{1}_a,$$

$$2° \quad \lambda \mathbf{1} \leq f_* \leq k \mathbf{1},$$

$$3° \quad \lambda \mathbf{1} \leq f^* \leq k \mathbf{1}.$$

Hence,

(16.5) For a subset $\{f_\alpha\}$ of C_a'', the following are equivalent:

$$1° \quad \{f_\alpha\} \text{ is bounded above,}$$

$$2° \quad \{(f_\alpha)_*\} \text{ is bounded above,}$$

$$3° \quad \{(f_\alpha)^*\} \text{ is bounded above.}$$

And similarly for "bounded below".

Remark. (16.4) gives us in particular that if one of f, f_*, f^* is positive, then the other two are also. One consequence: for $f \in (C_a'')_+$, $\mathbf{u}(f) = f^*$, $\boldsymbol{\ell}(f) = f$. (We already have the second of these (cf. (2.2)).)

(16.6) Given $f, g \in C_a''$,

$$f_* + g_* \leq (f+g)_* \leq f_* + g^* \leq (f+g)^* \leq f^* + g^*.$$

It follows

$$f_* - g^* \leq (f-g)_* \leq \begin{matrix} f^* - g^* \\[4pt] \\[4pt] f_* - g_* \end{matrix} \leq (f-g)^* \leq f^* - g_*.$$

(cf. (2.4))

(16.7) Given a bounded set $\{f_\alpha\}$ in C''_a,

(1) $(\wedge_\alpha f_\alpha)^* \le \wedge_\alpha (f_\alpha)^* \le \vee_\alpha (f_\alpha)^* \le (\vee_\alpha f_\alpha)^*.$

(2) $(\wedge_\alpha f_\alpha)_* \le \wedge_\alpha (f_\alpha)_* \le \vee_\alpha (f_\alpha)_* \le (\vee_\alpha f_\alpha)_*.$

(cf. (2.5))

For a countable set, the last inequality in (1) and the first in (2) become equalities:

(16.8) Given a countable bounded set $\{f_n\}$ in C''_a,

(1) $\vee_n (f_n)^* = (\vee_n f_n)^*,$

(2) $\wedge_n (f_n)_* = (\wedge_n f_n)_*.$

Proof. We prove (1). Set $f = \vee_n f_n$ and $h = \vee_n (f_n)^*$. We need only show $h \ge f^*$. Since the projection preserves suprema,

$$h_a = \vee_n [(f_n)^*]_a = \vee_n f_n = f,$$

so $h \in C''_d + f$. But $(f_n)^*$ is a **u**-element for every n, so by (2.1), h is also. Since f^* is the smallest **u**-element in $C''_d + f$, we have $h \ge f^*$.

QED

For a finite set in C''_a, we can sharpen some of our results.

(16.9) Given $f, g \in C''_a$,

(1) $(f \vee g)_* \le f_* \vee g^* \le (f \vee g)^*.$

(2) $(f \wedge g)_* \le f_* \wedge g^* \le (f \wedge g)^*.$

Proof. We show (1).

The first inequality: This follows from

$$(f \vee g)_* - f_* \leq f_* \vee g^* - f_*.$$

which we now establish.

$$(f \vee g)_* - f_* \leq (f \vee g - f)^* \qquad\qquad (16.6)$$
$$= [(g - f)^+]^*$$
$$= [(g - f)^*]^+ \qquad\qquad (16.8)$$
$$\leq (g^* - f_*)^+ \qquad\qquad (16.6)$$
$$= f_* \vee g^* - f_*.$$

The second inequality:

$$f_* \vee g^* \leq f^* \vee g^*$$
$$= (f \vee g)^*. \qquad\qquad (16.8)$$

$$\text{QED}$$

Combining the above with the Remark following (16.5), we obtain

(16.10) For $g, f \in C''_a$, the following are equivalent:

$1°$ $f \wedge g = 0$,

$2°$ $f^* \wedge g^* = 0$,

$3°$ $f_* \wedge g_* = 0$.

We sharpen the first and last inequalities in the first line of (16.6).

(16.11) Given $g, f \in C_a''$,

(1) $f_* + g_* \leq (f \vee g)_* + (f \wedge g)_* \leq (f + g)_*$.

(2) $(f + g)^* \leq (f \vee g)^* + (f \wedge g)^* \leq f^* + g^*$.

Proof. We prove (2).

$$(f+g)^* = (f \vee g + f \wedge g)^*$$

$$\leq (f \vee g)^* + (f \wedge g)^* \qquad (16.6)$$

$$= f^* \vee g^* + (f \wedge g)^* \qquad (16.8)$$

$$\leq f^* \vee g^* + f^* \wedge g^* \qquad (16.7)$$

$$= f^* + g^*.$$

QED

(16.12) Given a bounded sequence $\{f_n\}$ in C_a'',

(1) $\liminf_n (f_n)_* \leq [\liminf_n (f_n)]_*$

$$\leq [\liminf_n (f_n)]^*$$

$$\leq \liminf_n (f_n)^*.$$

(2) $\limsup_n (f_n)_* \leq [\limsup_n (f_n)]_*$

$$\leq [\limsup_n (f_n)]^*$$

$$\leq \limsup_n (f_n)^*.$$

It follows that if $f_n \longrightarrow f$, then

$$\limsup_n (f_n)_* \leq f_* \leq f^* \leq \liminf_n (f_n)^*.$$

(cf. the proof of (2.7))

We combine this with Fatou's Lemma (13.2).

(16.13) Given $\mu \in C'_+$ and a bounded sequence $\{f_n\}$ in C''_a,

$$\liminf_n \langle (f_n)_*, \mu \rangle.$$

(1) $\langle \liminf_n (f_n)_*, \mu \rangle \leq$

$$\langle [\liminf_n (f_n)]_*, \mu \rangle.$$

(2) $\langle [\liminf_n (f_n)]^*, \mu \rangle \leq \langle \liminf_n (f_n)^*, \mu \rangle \leq \liminf_n \langle (f_n)^*, \mu \rangle.$

(3) $\limsup_n \langle (f_n)_*, \mu \rangle \leq \langle \limsup_n (f_n)_*, \mu \rangle \leq \langle [\limsup_n (f_n)]_*, \mu \rangle.$

(4) $\limsup_n \langle (f_n)^*, \mu \rangle$

$$\leq \langle \limsup_n (f_n)^*, \mu \rangle.$$

$\langle [\limsup_n (f_n)]^*, \mu \rangle$

Given $f, g \in C''_a$, if one of them lies in U_a, then all our binomial inequalities become equalities:

(16.14) Given $g \in U_a$, then for every $f \in C''_a$:

(1) $f^* \vee g^* = (f \vee g)^*,$

(2) $f^* \wedge g^* = (f \wedge g)^*,$

(3) $f_* \vee g_* = (f \vee g)_*,$

(4) $f_* \wedge g_* = (f \wedge g)_*,$

(5) $f^* + g^* = (f + g)^*,$

(6) $f_* + g_* = (f + g)_*.$

Proof. (1) and (4) hold for every $f, g \in C_a''$ (16.8). Since $g_* = g^*$, (2) and (3) follow from (16.9), and (5) and (6) from (16.6).

QED

Setting $g = 0$, we obtain

(16.15) For every $f \in C_a''$:

$$(f^*)^+ = (f^+)^*,$$

$$(f^*)^- = (f^-)_*,$$

$$(f_*)^+ = (f^+)_*,$$

$$(f_*)^- = (f^-)^*.$$

Remark. From the identities obtained so far, it follows easily that for $f \in (C_a'')_+$,

$$[(\mathbf{1}_a)_f]^* = \mathbf{1}_{f^*}.$$

And more generally, for $f \in C_a''$ and $\lambda \geq 0$,

$$[(\mathbf{1}_a)_{(f - \lambda \mathbf{1}_a)^+}]^* = \mathbf{1}_{(f_* - \lambda \mathbf{1})^+}.$$

We will need these identities.

(16.16) Given $f \in C_a''$,

    (1) $|f|^* = |f^*| \vee |f_*|.$

    (2) $|f|_* \leq |f^*| \wedge |f_*|.$

Thus

$$|f|_* \leq \begin{matrix} |f^*| \\ \\ |f_*| \end{matrix} \leq |f|^*.$$

(cf. the proof of (2.11))

(16.17) For $f, g \in C_a''$,

$$\begin{matrix} |f^* - g^*| \\ \\ |f_* - g_*| \end{matrix} \leq |f-g|^*.$$

(cf. the proof of (2.12))

An easy consequence of (16.4) and the identity $\|f\| = \| |f| \|$ is that for $f \in C_a''$,

$$\|f\| = \| |f|^* \| = \max (\|f^*\|, \|f_*\|).$$

In particular, for $f \in (C_a'')_+$, $\|f\| = \|f^*\|$. Combining this with (16.17), we obtain

(16.18) For $f, g \in C_a''$,

$$\begin{matrix} \|f^* - g^*\| \\ \\ \|f_* - g_*\| \end{matrix} \leq \|f-g\|.$$

Thus the mappings $f \mapsto f_*$ and $f \mapsto f^*$ are norm continuous.

Finally,

(16.19) If $e \in C''_d$ is a component of $\mathbf{1}_a$ (equivalently, of $\mathbf{1}$), then e_* and e^* are components of $\mathbf{1}$.

$$\textit{Proof.} \quad (2e_*) \wedge \mathbf{1} = (2e)_* \wedge \mathbf{1} \qquad\qquad (16.3)$$
$$= [(2e) \wedge \mathbf{1}_a]_* \qquad\qquad (16.8)$$
$$= e_*.$$

Similarly for e^*.

QED

§17. $\mathfrak{L}^\infty$ and the star elements

The fibering of C'' by the translates of C''_d induces a fibering of $\mathfrak{L}^\infty$ by the translates of $\mathfrak{L}^\infty \cap C''_d$ (that is, translates of $\mathfrak{L}^\infty \cap C''_d$ by elements of $\mathfrak{L}^\infty$). Recall that $\mathfrak{L}^\infty \cap C''_d = \mathcal{N}_d$ (cf. the proof of (5.5)). Specifically,

(17.1) If a translate A of C''_d contains an element g of $\mathfrak{L}^\infty$, then
$$\mathfrak{L}^\infty \cap A = g + \mathcal{N}_d.$$
Otherwise stated, either $\mathfrak{L}^\infty$ does not intersect A or its intersection with A is a translate of $\mathcal{N}_d$ by an element of $\mathfrak{L}^\infty$.

Proof. That $g + \mathcal{N}_d \subset \mathfrak{L}^\infty \cap A$ is clear. For the opposite inclusion, consider $h \in \mathfrak{L}^\infty \cap A$. $h - g \in \mathfrak{L}^\infty$ and $h - g \in C_d''$, so $h - g \in \mathfrak{L}^\infty \cap C_d'' = \mathcal{N}_d$.

$$\text{QED}$$

(17.2) *Corollary.* Given a translate A of C_d'', if $g, h \in \mathfrak{L}^\infty \cap A$ with $g \le h$, then $[g, h] \subset \mathfrak{L}^\infty \cap A$.

Given $f \in C_a''$, if $\mathfrak{L}^\infty$ intersects $C_d'' + f$, then it contains f_* and f^*:

(17.3) For $f \in C_a''$, the following are equivalent:

    $1°$ $f \in (\mathfrak{L}^\infty)_a$,

    $2°$ $f^* \in \mathfrak{L}^\infty$,

    $3°$ $f_* \in \mathfrak{L}^\infty$,

    $4°$ $f^* - f_* \in \mathcal{N}$,

    $5°$ $f^* - f_* \in N$.

Proof. Assume $1°$; we show $2°$, $3°$, $4°$, and $5°$ hold. Choose $g \in \mathfrak{L}^\infty \cap (C_d'' + f)$. By (2.1) and (2.14), $\ell(g)$ and $u(g)$ lie in $\mathfrak{L}^\infty \cap (C_d'' + f)$. Now

$$\text{(i)} \quad \ell(g) \le f_* \le f^* \le u(g).$$

It follows from (17.2) that $2°$ and $3°$ hold. (i) also gives us

$$0 \le f^* - f_* \le u(g) - \ell(g) \in \mathcal{N},$$

so we have $4°$, hence also $5°$.

2° and 3° each imply 1° trivially. And finally, 5° implies 3°: in effect (cf. the discussion preceding (16.2)), $\delta(f^*) = f^* - f_* \in N$, so $f_* \in \mathfrak{L}^\infty$.

QED

(17.4) For $f \in C_a''$, the following are equivalent:

  1° $f \in N_a$,

  2° $f^* \in N$,

  3° $f_* \in N$,

  4° $f_*, f^* \in N$.

And in such case, $\mathfrak{L}^\infty \cap (C_d'' + f) \subset N$.

Proof. Assume 1°; we show 2°, 3°, and 4° hold. Choose $g \in N \cap (C_d'' + f)$. By (17.1),

$$\mathfrak{L}^\infty \cap (C_d'' + f) = g + N_d \subset N.$$

Combining this with (17.3) gives us that $f_*, f^* \in N$, so we have 2°, 3°, and 4°. 2° and 3° each imply 1°. Finally, 4° implies 3°: in effect, it implies 5° in (17.3), hence that $f_* \in \mathfrak{L}^\infty$, hence that $f^* \in \mathfrak{L}^\infty \cap N = N$.

QED

(17.3) and (17.4) give us

(17.5) Given $f \in C''_a$.

(1) The following are equivalent:

$1°$ $f \in (\mathfrak{L}^\infty)_a$,

$2°$ $\langle f_*, \nu \rangle = \langle f^*, \nu \rangle$ for every $\nu \in J$.

(2) And the following are equivalent:

$1°$ $f \in \mathcal{N}_a$,

$2°$ $\langle f_*, \nu \rangle = \langle f^*, \nu \rangle = 0$ for every $\nu \in J$.

(17.6) *Corollary.* Given $\mu \in C'_+$ and $f \in C''_a$.

(1) The following are equivalent:

$1°$ $f \in \mathfrak{L}^\infty(\mu)_a$,

$2°$ $\langle f_*, \mu \rangle = \langle f^*, \mu \rangle$.

(2) And the following are equivalent:

$1°$ $f \in \mathcal{N}(\mu)_a$,

$2°$ $\langle |f_*|, \mu \rangle = \langle |f^*|, \mu \rangle = 0$,

$3°$ $\langle |f|^*, \mu \rangle = 0$.

Proof. (1) That $1°$ implies $2°$ follows from (17.5). Assume $2°$. This can be written $\langle f^* - f_*, \mu \rangle = 0$. Since $f^* - f_* \geq 0$, it follows from (10.2) that $f^* - f_* \in N(\mu)$. Hence, by (17.3), $f_* \in \mathfrak{L}^\infty(\mu)$.

(2)   Assume $1°$.   By   (17.4), $f_*, f^* \in N(\mu)$,   hence $\langle |f_*|, \mu \rangle = \langle |f^*|, \mu \rangle = 0$, so $2°$ holds. Assume $2°$. then by (16.16),

$$0 \leq \langle |f|^*, \mu \rangle = \langle |f_*| \vee |f^*|, \mu \rangle \leq \langle |f_*| + |f^*|, \mu \rangle = 0,$$

so $3°$ holds. Assume $3°$. again by (16.16), $\langle |f_*|, \mu \rangle = \langle |f^*|, \mu \rangle = 0$.

Hence by (10.2), $f_*, f^* \in N(\mu)$. It follows from (17.4) that 1° holds.

QED

§18. $\mathfrak{L}_a^\infty$

Henceforth, we will write $\mathfrak{L}_a^\infty$ for $(\mathfrak{L}^\infty)_a$.

(18.1) $\mathfrak{L}_a^\infty$ is a σ-order closed Riesz subspace of C_a'' containing U_a.

Proof.   The   projection   of C'' onto C_a'' is   a   Riesz
homomorphism, so $\mathfrak{L}_a^\infty$ is a Riesz subspace of C_a''. Thus it suffices to
prove that if $\{f_n\} \subset \mathfrak{L}_a^\infty$ and $f = \vee_n f_n$, then $f \in \mathfrak{L}_a^\infty$. For each
$n = 1, 2, ...$, choose $h_n \in \mathfrak{L}^\infty$ such that $(h_n)_a = f_n$. We can take $\{h_n\}$ to
be bounded above: in effect, $\{f_n\} \leq \lambda \mathbf{1}_a$ for some $\lambda \in \mathbb{R}$; then the set
$\{h_n \wedge \lambda\mathbf{1}\}$ is bounded above and $(h_n \wedge \lambda\mathbf{1})_a = f_n$ for every n. Now let
$h = \vee_n h_n$. Then $h_a = f$, and since $h \in \mathfrak{L}^\infty$ (4.1), we have $f \in \mathfrak{L}_a^\infty$.

QED

Note that, unlike $\mathfrak{L}^\infty$, $\mathfrak{L}_a^\infty$ is, in general, not Dedekind closed
in C_a''. In effect, $\mathfrak{L}_a^\infty$ contains U_a, the Dedekind closure of which is
C_a'', so a fortiori, that of $\mathfrak{L}_a^\infty$ is, also. So we need only note that $\mathfrak{L}_a^\infty$
need not be all of C_a'' (for example, if X is a real interval and μ the
Lebesgue measure, then $\mathfrak{L}^\infty(\mu)_a$ is a proper subset of C_a'').

Since $\mathcal{N}$ is a Riesz ideal of C'', we have

(18.2) $\mathcal{N}_a = \mathcal{N} \cap C_a''$, hence is a σ-order closed Riesz ideal of C_a''.

(18.3) (1) $\mathfrak{L}_a^\infty = \cap_{\nu \in J_+} \mathfrak{L}^\infty(\nu)_a$.

(2) $\mathcal{N}_a = \cap_{\nu \in J_+} \mathcal{N}(\nu)_a$.

Proof. (1) For every $\nu \in J_+$, $\mathfrak{L}^\infty \subset \mathfrak{L}^\infty(\nu)$ (6.5), so $\mathfrak{L}_a^\infty \subset \mathfrak{L}^\infty(\nu)_a$. Thus $\mathfrak{L}_a^\infty \subset \cap_{\nu \in J_+} \mathfrak{L}^\infty(\nu)_a$. For the opposite inclusion, suppose $f \in \cap_{\nu \in J_+} \mathfrak{L}^\infty(\nu)_a$. So for every $\nu \in J_+$, $f \in \mathfrak{L}^\infty(\nu)_a$, hence, by (17.3), $f^* \in \mathfrak{L}^\infty(\nu)$. Thus $f^* \in \cap_{\nu \in J_+} \mathfrak{L}^\infty(\nu) = \mathfrak{L}^\infty$ (again by (6.5)), hence $f \in \mathfrak{L}_a^\infty$.

(2) follows from (6.5) and (18.2) above.   QED

If E and F are normed Riesz spaces with the norms given by strong order units $\mathbf{1}(E)$ and $\mathbf{1}(F)$ respectively, then the notation $E \cong F$ will mean that E and F are Riesz isomorphic with $\mathbf{1}(E) \longleftrightarrow \mathbf{1}(F)$. It follows the isomorphism is also an isometry.

$$\mathfrak{L}_a^\infty \cong \mathfrak{L}^\infty / \mathfrak{L}^\infty \cap C_d'' = \mathfrak{L}^\infty / \mathcal{N}_d.$$

This isomorphism induces an isometric Riesz isomorphism $\mathcal{N}_a \cong \mathcal{N}/\mathcal{N}_d$. So

$$\mathfrak{L}_a^\infty / \mathcal{N}_a \cong (\mathfrak{L}^\infty / \mathcal{N}_d)/(\mathcal{N}/\mathcal{N}_d) \cong \mathfrak{L}^\infty / \mathcal{N}.$$

Moreover, since $C'' = I \oplus N$ and $\mathcal{N} = \mathfrak{L}^\infty \cap N$, we have

$$(\mathfrak{L}^\infty)_I \cong \mathfrak{L}^\infty / \mathcal{N}.$$

This gives us most of

(18.4) $\mathfrak{L}^{\infty}/\mathcal{N} \cong \mathfrak{L}_a^{\infty}/\mathcal{N}_a \cong (\mathfrak{L}^{\infty})_{\mathrm{I}}.$

And under these isomorphisms, the following correspond:

$1°$ the image of $\mathbf{1}$ in $\mathfrak{L}^{\infty}/\mathcal{N}$,

$2°$ the image of $\mathbf{1}_a$ in $\mathfrak{L}_a^{\infty}/\mathcal{N}_a$,

$3°$ $\mathbf{1}_{\mathrm{I}}$.

§19. Convergence in C_a''

We note first that as far as the almost everywhere properties in which we are interested are concerned, for elements of C_a'', $\mathcal{N}$ and $\mathcal{N}_a$ are interchangeable. Specifically:

(1) For $f, g \in C_a''$, the following are equivalent:

$1°$ $g \leq f$ a.e. in C_a'' (with respect to $\mathcal{N}_a$),

$2°$ $g \leq f$ a.e. in C'' (with respect to $\mathcal{N}$).

(2) For $\{f_n\} \subset C_a''$ and $h \in C''$, the following are equivalent:

$1°$ $h = \vee_n f_n$ a.e. in C_a'' (with respect to $\mathcal{N}_a$),

$2°$ $h = \vee_n f_n$ a.e. in C'' (with respect to $\mathcal{N}$).

(3) For $\{f_n\} \subset C_a''$ and $f \in C_a''$, the following are equivalent:

$1°$ $f_n \xrightarrow{u} f$ a.e. in C_a'' (with respect to $\mathcal{N}_a$),

$2°$ $f_n \xrightarrow{u} f$ a.e. in C'' (with respect to $\mathcal{N}$).

We prove (1). Assume $1°$. So there exists a band H of C_a'' contained in $\mathcal{N}_a$ such that

$$g_{(\mathrm{H}^d \cap C_a'')} \leq f_{(\mathrm{H}^d \cap C_a'')}.$$

Now $f, g \in C_a''$, so this can be written

$$g_{H^d} \leq f_{H^d}.$$

Since H is also a band of C'' and $\mathcal{N}_a \subset \mathcal{N}$, we have $2°$.

Assume $2°$. So there exists a band H of C'' contained in $\mathcal{N}$ such that $g_{H^d} \leq f_{H^d}$. Now H_a is a band of C_a'' contained in $\mathcal{N}_a$, and $(H_a)^d \cap C_a'' = H^d \cap C_a''$, so

$$g_{[(H_a)^d \cap C_a'']} = g_{(H^d \cap C_a'')}$$

$$= g_{H^d}$$

$$\leq f_{H^d}$$

$$= f_{(H^d \cap C_a'')}$$

$$= f_{[(H_a)^d \cap C_a'']}.$$

We thus have $1°$.

It follows from the above, that for countable subsets of C_a'' and sequences in C_a'', we can make almost everywhere statements without specifying whether they hold in C_a'' with respect to $\mathcal{N}_a$ or in C'' with respect to $\mathcal{N}$.

We sharpen (1) in (16.3).

(19.1) For $f, g \in C_a''$, the following are equivalent:

$1°$ $g \leq f$ a.e.,

$2°$ $g^* \leq f^*$ a.e.,

$3°$ $g_* \leq f_*$ a.e..

Proof. It is convenient to prove the equivalence of these written in the following form (cf. the Scholium at the end of §7):

1° $g \leq f \mod \mathcal{N}_a$,

2° $g^* \leq f^* \mod \mathcal{N}$,

3° $g_* \leq f_* \mod \mathcal{N}$.

Write them

1° $(g-f)^+ \in \mathcal{N}_a$,

2° $(g^*-f^*)^+ \in \mathcal{N}$,

3° $(g_*-f_*)^+ \in \mathcal{N}$.

Note first that by (16.15) and (16.16),

$$(g^*-f^*)^+$$

(i) $[(g-f)^+]_* \leq \leq [(g-f)^+]^*.$

$$(g_*-f_*)^+$$

Now assume 1°. By (17.4), $[(g-f)^+]_*$ and $[(g-f)^+]^*$ lie in $\mathcal{N}$; hence by (i), $(g^*-f^*)^+$ and $(g_*-f_*)^+$ do also. We thus have 2° and 3°. Assume 2°. It follows from (i) that $[(g-f)^+]_* \in \mathcal{N}$, hence $(g-f)^+ \in \mathcal{N}_a$. So 1° holds. That 3° implies 1° is proved similarly.

$$\text{QED}$$

Combining this with (7.8) gives us

(19.2) *Corollary.* For $\{f_n\} \subset C''_a$ and $h \in C''_a$, the following are equivalent:

1° $\{f_n\}$ is bounded above by h a.e.,

2° $\{(f_n)^*\}$ is bounded above by h^* a.e.,

$3°$ $\{(f_n)_*\}$ is bounded above by h_* a.e..

And similarly with "above" replaced by "below".

(19.3) (1) For $\{f_n\} \subset C_a''$ and $h \in C_a''$, the following are equivalent:

   $1°$ $h = \vee_n f_n$ a.e.,

   $2°$ $h^* = \vee_n (f_n)^*$ a.e..

   (2) And the following are equivalent:

   $1°$ $h = \wedge_n f_n$ a.e.,

   $2°$ $h_* = \wedge_n (f_n)_*$ a.e..

Proof. We prove (1). That $2°$ implies $1°$ follows from (7.11). Assume $1°$. Set $g_n = f_n \wedge h$ $(n=1,2,...)$ and $g = \vee_n g_n$. By (16.8), we have

   (i) $g^* = \vee_n (g_n)^*$.

We show:

   (a) $(f_n)^* = (g_n)^*$ a.e. for every n,

   (b) $h^* = g^*$ a.e..

It will follow from (7.10) and (i) that $2°$ holds. By (7.4), $f_n = g_n$ a.e. for every n, so (19.1) gives us (a). And by (7.10), $h = g$ a.e., so (19.1) also gives us (b).

                                                                QED

It follows from (19.1)–(19.3) that an ascending sequence $\{f_n\}$ in C_a'' order converges, or order converges a.e., to $f \in C_a''$ if and only if $\{(f_n)^*\}$ converges to f^* in the same mode. And similarly for a descending sequence with the upper stars replaced by lower stars. Except for these, we have no theorems in which the convergence of a sequence in C''_a in one of the above two modes implies the convergence of its upper star or lower star elements. We will see in §20 that for sequences in $\mathfrak{L}_a^\infty$, we have all that we could desire.

We turn to convergence in measure (note the lower case m).

We remark first that an element f of C''_a often cannot distinguish between $\mathbf{1}$ and $\mathbf{1}_a$. For example, $f \wedge \mathbf{1} = f \wedge \mathbf{1}_a$, $\mathbf{1}_f = (\mathbf{1}_a)_f$, and for $\lambda \geq 0$, $(f - \lambda\mathbf{1})^+ = (f - \lambda\mathbf{1}_a)^+$. Consequently, we will feel free to use the above left expressions in place of those on the right when convenient.

(19.4)   Given $\mu \in C'_+$, then for a sequence $\{f_n\}$ in $(C''_a)_+$, the following are equivalent:

$$1° \lim_n \langle (f_n \wedge \mathbf{1}_a)^*, \mu \rangle = 0,$$

$$2° \lim_n \langle [(\mathbf{1}_a)_{(f_n - \lambda\mathbf{1}_a)^+}]^*, \mu \rangle = 0 \text{ for every } \lambda > 0.$$

Proof.

Lemma. For every $g \in C''_a$ and $\lambda \in \mathbb{R}$:

(1) $[(\mathbf{1}_a)_g]^* = \mathbf{1}_{g^*}$,

(2) $[(\mathbf{1}_a)_{(g - \lambda\mathbf{1}_a)^+}]^* = \mathbf{1}_{(g^* - \lambda\mathbf{1})^+}.$

These can be verified by straightforward computation using the above discussion together with (16.14) and (16.15).

By (16.14) and the Lemma, 1° and 2° can be written

1° $\lim_n \langle (f_n)^* \wedge \mathbf{1}, \mu \rangle = 0,$

2° $\lim_n \langle \mathbf{1}_{[(f_n)^* - \lambda 1]^+}, \mu \rangle = 0$ for every $\lambda > 0.$

Hence, by (11.2), they are equivalent.   QED

If one, hence also the other, of the above statements holds, we will say that $\{f_n\}$ converges to 0 *in μ-measure* (lower case m), and we will write

$$f_n \longrightarrow 0 \text{ in } \mu\text{-measure.}$$

Thus, by very definition, for a sequence $\{f_n\}$ in $(C_a'')_+$, $f_n \longrightarrow 0$ in μ-measure if and only if $(f_n)^* \longrightarrow 0$ in μ-Measure.

More generally, we will say a sequence $\{f_n\}$ in C_a'' converges to 0 *in μ-measure* if $\{ |f_n| \}$ does.

(19.5) Given $\mu \in C_+'$, then for every $\{f_n\} \subset C_a''$, the following are equivalent:

1° $f_n \longrightarrow 0$ in μ-measure;

2° (i) $(f_n)^* \longrightarrow 0$ in μ-Measure,

(ii) $(f_n)_* \longrightarrow 0$ in μ-Measure.

Proof. Write 1° and 2° in the form:

1° $\lim_n \langle |f_n|^* \wedge \mathbf{1}, \mu \rangle = 0;$

$2°$ (i) $\lim_n \langle |(f_n)^*| \wedge \mathbf{1}, \mu \rangle = 0,$

(ii) $\lim_n \langle |(f_n)_*| \wedge \mathbf{1}, \mu \rangle = 0.$

Their equivalence then follows from (16.16).

QED

Given $\mu \in C'_+$, we will say a sequence $\{f_n\}$ in C''_a converges *in* μ-*measure* to $f \in C''_a$ — in notation, $f_n \longrightarrow f$ in μ-measure — if $f_n - f \longrightarrow 0$ in μ-measure.

Half of (19.5) carries over to this general case:

(19.6) Given $\mu \in C'_+$, if $\{f_n\} \subset C''_a$, $f \in C''_a$, and $f_n \longrightarrow f$ in μ-measure, then:

(1) $(f_n)^* \longrightarrow f^*$ in μ-Measure,

(2) $(f_n)_* \longrightarrow f_*$ in μ-Measure.

This follows from (16.17).

We do not know whether the converse implication holds. But we will see later that the converse does hold if $\{f_n\} \subset \mathfrak{L}_a^\infty$.

We turn to our general band J of C'. We will write (for $\{f_n\} \subset C''_a$) $f_n \longrightarrow f$ *in J-measure*, or simply *in measure*, if $f_n \longrightarrow f$ in ν-measure for every $\nu \in J_+$.

(11.6) gives us

(19.7) Given $\mu \in C'_+$, then for a sequence $\{f_n\}$ in C''_a and $f \in C''_a$, the following are equivalent:

$1°$ $f_n \longrightarrow f$ in C'_μ-measure,

$2°$ $f_n \longrightarrow f$ in μ-measure.

Again, it follows easily from this that any proposition on convergence in measure (that is, in J-measure) holds also for convergence in μ-measure, and conversely.

It is clear that for a sequence in C''_a bounded a.e., the statement "$f_n \longrightarrow f$ in μ-measure" reduces to "$\lim_n \| \, |f_n - f|^* \, \| = 0$" and the statement "$f_n \longrightarrow f$ in J-measure" reduces to "$\{ \, |f_n - f|^* \}$ $|\sigma| \, (C'', J)$-converges to 0" (cf. (16.5), (11.4), and (11.5)).

(19.8) Given $\{f_n\}$, $\{g_n\} \subset C''_a$ and $f, g \in C''_a$:

(1) If $f_n \longrightarrow f$ in measure, then $\rho f_n \longrightarrow \rho f$ in measure for every $\rho \in \mathbb{R}$.

(2) If $f_n \longrightarrow f$ in measure and $g = f$ a.e., then $f_n \longrightarrow g$ in measure.

(3) If $f_n \longrightarrow f$ in measure and $g_n = f_n$ a.e. for $n = 1, 2, \ldots$, then $g_n \longrightarrow f$ in measure.

(4) If $f_n \longrightarrow f$ in measure and $g_n \longrightarrow g$ in measure, then:

(i) $f_n + g_n \longrightarrow f + g$ in measure,

(ii) $f_n \vee g_n \longrightarrow f \vee g$ in measure,

(iii) $f_n \wedge g_n \longrightarrow f \wedge g$ in measure.

Proof. (1) follows from (11.8) and (16.3). We prove (2). Fix $\nu \in J_+$.

$$
\begin{aligned}
\langle \, | \, f_n - g \, |^* \wedge \mathbf{1}, \nu \rangle &\leq \langle ((\, | \, f_n - f \, | + | \, f - g \, |)^* \wedge \mathbf{1}, \nu \rangle \\
&\leq \langle ((\, | \, f_n - f \, |^* + | \, f - g \, |^*) \wedge \mathbf{1}, \nu \rangle \qquad (16.6) \\
&\leq \langle \, | \, f_n - f \, |^* \wedge \mathbf{1} + | \, f - g \, |^* \wedge \mathbf{1}, \nu \rangle \\
&= \langle \, | \, f_n - f \, |^* \wedge \mathbf{1}, \nu \rangle + \langle \, | \, f - g \, |^* \wedge \mathbf{1}, \nu \rangle .
\end{aligned}
$$

In the last expression, the first term converges to 0 by hypothesis. The second term is 0: in effect, $f - g \in \mathcal{N}_a$, so $| \, f - g \, | \in \mathcal{N}_a$, so (by (17.4)) $| \, f - g \, |^* \in \mathcal{N}$, so $| \, f - g \, |^* \wedge \mathbf{1} \in \mathcal{N}$. We leave the verification of the remaining properties to the reader.

$$\text{QED}$$

One consequence of the above:

(19.9)   For a sequence $\{ f_n \}$ in C_a'' and $f \in C_a''$, the following are equivalent:

$1°$ $f_n \longrightarrow f$ in measure;

$2°$ (i) $(f_n)^+ \longrightarrow f^+$ in measure,

    (ii) $(f_n)^- \longrightarrow f^-$ in measure.

And they imply

$$| \, f_n \, | \longrightarrow | \, f \, | \text{ in measure.}$$

Moreover, if $f = 0$, then the last statement is equivalent to the first two.

§20. Back to $\mathfrak{L}_a^\infty$

We sharpen (18.1).

(20.1) $\mathfrak{L}_a^\infty$ is closed in C_a'' under convergence of sequences in each of the following modes:

    (1) unbounded order convergence,

    (2) unbounded order convergence a.e.,

    (3) convergence in measure.

Proof. That $\mathfrak{L}_a^\infty$ is closed under mode (2), hence also under mode (1), is proved by the same kind of argument as was used for (12.1). We show it is closed under convergence of sequences in mode (3).

Assume $\{f_n\} \subset \mathfrak{L}_a^\infty$, $f \in C_a''$, and $f_n \longrightarrow f$ in measure. We show first that $\{f_n\}$ can be replaced by a bounded sequence in $\mathfrak{L}_a^\infty$. Choose $m \in \mathbb{N}$ such that $-m\mathbf{1}_a \leq f \leq m\mathbf{1}_a$, and set

$$g_n = (f_n \wedge m\mathbf{1}_a) \vee (-m\mathbf{1}_a) \quad (n=1, 2, \ldots).$$

$\{g_n\} \subset \mathfrak{L}_a^\infty$; we show $g_n \longrightarrow f$ in measure. As is well known, $|g_n - f| \leq |f_n - f|$, hence (cf. (16.3)) $|g_n - f|^* \wedge \mathbf{1} \leq |f_n - f|^* \wedge \mathbf{1}$, hence

$$0 \leq \langle |g_n - f|^* \wedge \mathbf{1}, \nu \rangle \leq \langle |f_n - f|^* \wedge \mathbf{1}, \nu \rangle \text{ for every } \nu \in J_+.$$

It follows $g_n \longrightarrow f$ in measure.

So from now on we assume $\{f_n\}$ is bounded. But then, by the comment preceding (19.8), the hypothesis can be written

    (i) $\lim_n \langle |f_n - f|^*, \nu \rangle = 0$ for every $\nu \in J_+$.

We complete the proof by showing that $f^* - f_* \in N$ (hence, by (17.3), $f \in \mathfrak{L}_a^\infty$).

Fix $\nu \in J_+$. By (i) and (16.17), we have

(ii) $\lim_n \langle |f_n)^* - f^*|, \nu \rangle = 0$,

(iii) $\lim_n \langle |f_n)_* - f_*|, \nu \rangle = 0$.

Now for every n,

$$0 \leq \langle |f^* - f_*|, \nu \rangle$$

$$\leq \langle |f^* - (f_n)^*|, \nu \rangle + \langle |(f_n)^* - (f_n)_*|, \nu \rangle + \langle |(f_n)_* - f_*|, \nu \rangle.$$

Since $f_n \in \mathfrak{L}_a^\infty$, the middle term in the last expression is 0. Hence by (ii) and (iii), $\langle |f^* - f_*|, \nu \rangle = 0$. Since this holds for every $\nu \in J_+$, $f^* - f_* \in N$.

QED

Likewise,

(20.2) N_a is closed in C_a'' under convergence of sequences in each of the following modes:

(1) unbounded order convergence,

(2) unbounded order convergence a.e.,

(3) convergence in measure.

Proof. We prove (2) (hence also (1)). Suppose $\{f_n\} \in N_a$, $f \in C_a''$, and $f_n \xrightarrow{u} f$ a.e.. $N_a = N \cap C_a''$, so $\{f_n\} \subset N$. Hence by (12.1) (cf. the discussion at the beginning of §19), $f \in N$. Thus $f \in N \cap C_a'' = N_a$. The proof that N_a is closed under convergence of sequences in mode (3) is similar to the proof for (20.1) above. We leave the details to the reader.

QED

Remark. Note that in contrast to the above $\mathfrak{L}$ and $\mathcal{N}$ are *not* closed under convergence in Measure of sequences (cf. the Remark following (12.1)).

The relations established in §§16, 19 of the elements of C_a'' to their star elements hold, of course, for the elements of $\mathfrak{L}_a^\infty$. But for the latter, we can say much more. We have shown some of this in §17. In particular, (from (17.3)), $f \in \mathfrak{L}_a^\infty$ if and only if $f_* = f^*$ a.e.. This enables us to strengthen, for $\mathfrak{L}_a^\infty$, many of the almost everywhere statements in §16.

From (2) and (3) of (16.3):

(20.3) For every $f \in \mathfrak{L}_a^\infty$ and $\rho \in \mathbb{R}$,

$$(\rho f)^* = (\rho f)_* = \rho f^* = \rho f_* \text{ a.e.}.$$

(We remind the reader that the "a.e." at the end applies to all the equalities.)

From (16.6),

(20.4) For $f, g \in \mathfrak{L}_a^\infty$:

$$f_* + g_* = (f+g)_* = (f+g)^* = f^* + g^* \text{ a.e.,}$$

$$f_* - g_* = (f-g)_* = (f-g)^* = f^* - g^* \text{ a.e.. }$$

From (16.7) and (16.8):

(20.5) For a countable bounded set $\{f_n\}$ in $\mathfrak{L}_a^\infty$:

$$\vee_n (f_n)^* = (\vee_n f_n)^*,$$

$$\wedge_n (f_n)^* = (\wedge_n f_n)^* \text{ a.e.,}$$

$$\vee_n (f_n)_* = (\vee_n f_n)_* \text{ a.e.,}$$

$$\wedge_n (f_n)_* = (\wedge_n f_n)_*.$$

In particular,

(20.6) For $f, g \in \mathfrak{L}_a^\infty$:

$$f^* \vee g^* = (f \vee g)^*,$$

$$f^* \wedge g^* = (f \wedge g)^* \text{ a.e.,}$$

$$f_* \vee g_* = (f \vee g)_* \text{ a.e.,}$$

$$f_* \wedge g_* = (f \wedge g)_*.$$

(20.7) *Corollary.* For $f, g \in \mathfrak{L}_a^\infty$, the following are equivalent:

$1°$ $f \wedge g = 0$ a.e.,

$2°$ $f^* \wedge g^* = 0$ a.e.,

$3°$ $f_* \wedge g_* = 0$ a.e..

From (16.15),

(20.8) Given $f \in \mathfrak{L}_a^\infty$.

$$(f^*)^+ = (f^+)^*.$$

$$(f_*)^+ = (f^+)_*.$$

$$(f^*)^- = (f^-)^* \text{ a.e. .}$$

$$(f_*)^- = (f^-)_* \text{ a.e. .}$$

One consequence:

$$|f^*| = |f|^* = |f_*| = |f|_* \text{ a.e. .}$$

Another (combining the above with (20.4)):

(20.9) Given $f, g \in \mathfrak{L}_a^\infty$.

$$|f^* + g^*| = |f + g|^* \text{ a.e. .}$$

$$|f^* - g^*| = |f - g|^* \text{ a.e. .}$$

Given $\{f_n\} \subset C''$ and $f \in C''$, we will write

$$f = \operatorname*{limsup}_n (f_n) \text{ a.e.}$$

if there exists a band H of C'' contained in $\mathcal{N}$ such that $f_{H^d} = \operatorname{limsup}_n (f_n)_{H^d}$. And similar for

$$f = \operatorname*{liminf}_n (f_n) \text{ a.e. .}$$

Note that the first definition implies that $\{f_n\}$ is bounded above a.e., and the second that it is bounded below a.e. .

Suppose $f = \operatorname*{limsup}_n (f_n)$ a.e. . It is easily verified that if $g = f$

a.e., then $g = \limsup_n (f_n)$ a.e., and conversely. Also, if $g_n = f_n$ a.e. $(n=1,2,...)$, then $f = \limsup_n (g_n)$ a.e.. Note further that if $\{f_n\} \subset \mathfrak{L}_a^\infty$ and $f = \liminf_n (f_n)$ a.e. (respectively, $\limsup_n (f_n)$ a.e.), then $f \in \mathfrak{L}_a^\infty$.

It follows from (16.12) (cf. also (19.3)) that

(20.10) Given a sequence $\{f_n\}$ in $\mathfrak{L}_a^\infty$, suppose $g = \liminf_n \{f_n\}$ a.e. and $h = \limsup_n (f_n)$ a.e.. Then

$$g^* = \liminf_n (f_n)^* \text{ a.e.,}$$

$$h^* = \limsup_n (f_n)^* \text{ a.e..}$$

And anywhere in these statements, the upper star can be replaced by a lower star.

(20.11) *Corollary.* For a sequence $\{f_n\}$ in $\mathfrak{L}_a^\infty$, the following are equivalent:

$1°$ $f_n \longrightarrow f$ a.e.,

$2°$ $(f_n)^* \longrightarrow f^*$ a.e.,

$3°$ $(f_n)_* \longrightarrow f_*$ a.e..

From this, we can easily obtain the more general

(20.12) For a sequence $\{f_n\}$ in $\mathfrak{L}_a^\infty$, the following are equivalent:

$1°$ $f_n \xrightarrow{u} f$ a.e.,

$2°$ $(f_n)^* \xrightarrow{u} f^*$ a.e.,

$3°$ $(f_n)_* \xrightarrow{u} f_*$ a.e..

(20.13) *Corollary.* For a sequence $\{f_n\}$ in $\mathfrak{L}_a^\infty$, $f_n \xrightarrow{u} f$ a.e. implies $f_n \rightarrow f$ in measure.

Proof.   By (20.12), $(f_n)^* \xrightarrow{u} f^*$ a.e., that is, $|(f_n)^* - f^*| \wedge \mathbf{1} \rightarrow 0$ a.e.. By (20.9), this can be written $|f_n - f|^* \wedge \mathbf{1} \rightarrow 0$ a.e., hence (cf. the remark following (10.1)),

$$\lim_n \langle |f_n - f|^* \wedge \mathbf{1}, \nu \rangle = 0 \ \text{ for every } \nu \in J_+.$$

Finally, by (20.6), this can be written

$$\lim_n \langle (|f_n - f| \wedge \mathbf{1}_a)^*, \nu \rangle = 0 \ \text{ for every } \nu \in J_+.$$

Thus $f_n \rightarrow f$ in measure.

QED

We can also show that for a sequence in $\mathfrak{L}_a^\infty$, not only (19.6) but also its converse hold. Indeed, we have a stronger result:

(20.14) For a sequence $\{f_n\}$ in $\mathfrak{L}_a^\infty$, the following are equivalent:

$1°$ $f_n \rightarrow f$ in measure,

$2°$ $(f_n)^* \rightarrow f^*$ in Measure,

$3°$ $(f_n)_* \rightarrow f_*$ in Measure.

Proof. The statements can be written

$1°$ $\lim_n \langle |f_n - f|^* \wedge \mathbf{1}, \nu \rangle = 0$ for every $\nu \in J_+$,

$2°$ $\lim_n \langle \, | \, (f_n)^* - f^* \, | \, \wedge \mathbf{1}, \nu \rangle = 0$ for every $\nu \in J_+$,

$3°$ $\lim_n \langle \, | \, (f_n)_* - f_* \, | \, \wedge \mathbf{1}, \nu \rangle = 0$ for every $\nu \in J_+$.

So the equivalence follows from (20.9).

QED

Chapter 5. Lebesgue Theory in C_a''

We have identified C_a'' with $\ell^\infty(X)$, so by the title, we mean standard Lebesgue theory for the bounded functions on X (with respect to the Radon — equivalently, regular — measures on X). We show that for each $\mu \in C'_+$, $\mathfrak{L}^\infty(\mu)_a$ is the space of bounded μ-integrable functions on X; and more generally, for our band J of C', $\mathfrak{L}_a^\infty$ is the space of bounded functions on X ν-integrable for every $\nu \in J_+$. We then show that under the projection onto C_a'', the theorems in Chapter 3 give us the corresponding theorems in standard Lebesgue theory.

$$\S 21. \quad \int^* fd\mu, \quad \int_* fd\mu$$

Given $\mu \in C'_+$, then for each $f \in C_a''$, we denote the value $\langle f^*, \mu \rangle$ by $\int^* fd\mu$, and the value $\langle f_*, \mu \rangle$ by $\int_* fd\mu$. We will call the first of these the *upper (Darboux) integral of* f *with respect to* μ, or simply the *upper μ-integral of* f; and similarly for the second, with "upper" replaced by "lower".

As we know, $f \in U_a$ if and only if $f_* = f^*$. This last is equivalent to $\langle f_*, \mu \rangle = \langle f^*, \mu \rangle$ for every $\mu \in C'_+$. We thus have

(21.1) For $f \in C''_a$, the following are equivalent:

$1°$ $f \in U_a$,

$2°$ $\int_* f d\mu = \int^* f d\mu$ for every $\mu \in C'_+$.

We now fix $\mu \in C'_+$ for the remainder of this §.

(16.2) gives us

(21.2) Given $f \in C''_a$.

(1) $\int^* f d\mu = \inf \{\langle h, \mu \rangle : h \in U, h_a \geq f\}$

$= \inf \{\langle h, \mu \rangle : h$ an lsc element, $h_a \geq f\}$.

(2) $\int_* f d\mu = \sup \{\langle g, \mu \rangle : g \in U, g_a \leq f\}$

$= \sup \{\langle g, \mu \rangle : g$ a usc element, $g_a \leq f\}$.

Proof. (1) The set $\{h \in U: h_a \geq f\}$ can be considered a descending net order converging to f^*. Since μ is order continuous on C'', this gives us the first equality. The remainder of (1) and the two equalities of (2) are proved similarly.

QED

(21.3) Given $f \in C''_a$.

(1) $\int_* f d\mu \leq \int^* f d\mu$.

(2) $\int^* (-f) d\mu = - \int_* f d\mu$.

(3) For every $\rho \in \mathbb{R}_+$:

$$\int {}^*\!\rho\mathrm{f}\mathrm{d}\mu = \rho \int {}^*\!\mathrm{f}\mathrm{d}\mu,$$

$$\int {}_*\!\rho\mathrm{f}\mathrm{d}\mu = \rho \int {}_*\!\mathrm{f}\mathrm{d}\mu.$$

It is easily seen that for $\mathrm{h}, \mathrm{k} \in C''$, $\mathrm{h} \leq \mathrm{k}$, μ-a.e. implies $\langle \mathrm{h}, \mu \rangle \leq \langle \mathrm{k}, \mu \rangle$. Hence (19.1) gives us

(21.4) Given $\mathrm{f}, \mathrm{g} \in C''_a$, if $\mathrm{g} \leq \mathrm{f}$ μ-a.e., then:

$$\int {}^*\!\mathrm{g}\mathrm{d}\mu \leq \int {}^*\!\mathrm{f}\mathrm{d}\mu,$$

$$\int {}_*\!\mathrm{g}\mathrm{d}\mu \leq \int {}_*\!\mathrm{f}\mathrm{d}\mu.$$

In particular, $\mathrm{f} \geq 0$ μ-a.e. implies $\int {}_*\!\mathrm{f}\mathrm{d}\mu \geq 0$ and $\int {}^*\!\mathrm{f}\mathrm{d}\mu \geq 0$.

From (16.4) and the identity $\langle \mathbf{1}, \mu \rangle = \| \mu \|$, we have

(21.5) If $\lambda \mathbf{1}_a \leq \mathrm{f} \leq \kappa \mathbf{1}_a$, then

$$\lambda \| \mu \| \leq \int {}_*\!\mathrm{f}\mathrm{d}\mu \leq \int {}^*\!\mathrm{f}\mathrm{d}\mu \leq \kappa \| \mu \|.$$

From (16.6),

(21.6) (1) Given $\mathrm{f}, \mathrm{g} \in C''_a$.

$$\int {}_*\!\mathrm{f}\mathrm{d}\mu + \int {}_*\!\mathrm{g}\mathrm{d}\mu \leq \int {}_*\!(\mathrm{f}+\mathrm{g})\mathrm{d}\mu$$

$$\leq \int {}_*\!\mathrm{f}\mathrm{d}\mu + \int {}^*\!\mathrm{g}\mathrm{d}\mu$$

$$\leq \int {}^*\!(\mathrm{f}+\mathrm{g})\mathrm{d}\mu$$

$$\leq \int {}^*\!\mathrm{f}\mathrm{d}\mu + \int {}^*\!\mathrm{g}\mathrm{d}\mu.$$

Hence

(2) $\int_* f d\mu - \int^* g d\mu \leq \int_* (f-g) d\mu$

$$\leq \begin{cases} \int^* f d\mu - \int^* g d\mu \\ \int_* f d\mu - \int_* g d\mu \end{cases}$$

$$\leq \int^* (f-g) d\mu$$

$$\leq \int^* f d\mu - \int_* g d\mu.$$

We sharpen the first and last inequalities in (1).

(21.7) Given $f, g \in C_a''$.

   (a) $\int_* f d\mu + \int_* g d\mu \leq \int_* (f \vee g) d\mu + \int_* (f \wedge g) d\mu \leq \int_* (f+g) d\mu.$

   (b) $\int^* (f+g) d\mu \leq \int^* (f \vee g) d\mu + \int^* (f \wedge g) d\mu \leq \int^* f d\mu + \int^* g d\mu.$

This follows from (16.11).

(21.8) Given $\{f_\alpha\} \subset C_a''$, $g = \wedge_\alpha f_\alpha$ μ-a.e., and $h = \vee_\alpha f_\alpha$ μ-a.e..

   (1) $\int^* g d\mu \leq \inf_\alpha \int^* f_\alpha d\mu \leq \sup_\alpha \int^* f_\alpha d\mu \leq \int^* h d\mu.$

   (2) $\int_* g d\mu \leq \inf_\alpha \int_* f_\alpha d\mu \leq \sup_\alpha \int_* f_\alpha d\mu \leq \int_* h d\mu.$

Proof. We prove the first inequality in (1). For each α, $g \leq f_\alpha$ μ-a.e. (cf. (7.9)), hence by (21.4), $\int^* g d\mu \leq \int^* f_\alpha d\mu$. It follows $\int^* g d\mu \leq \inf_\alpha \int^* f_\alpha d\mu$.

                                                          QED

From (2) and (3) of (16.13), we have

(21.9) (Fatou) Given a bounded sequence $\{f_n\}$ in C''_a.

$$\int {}^* \liminf_n (f_n) \, d\mu \ \leq \ \liminf_n \int {}^* f_n d\mu.$$

$$\limsup_n \int_* f_n d\mu \ \leq \ \int_* \limsup_n (f_n) d\mu.$$

(21.10) *Corollary.* given a bounded sequence $\{f_n\}$ in C''_a, if $f_n \longrightarrow f$, then

$$\limsup_n \int_* f_n d\mu \ \leq \ \int_* f d\mu \ \leq \ \int {}^* f d\mu \ \leq \ \liminf_n \int {}^* f_n d\mu.$$

(21.11) The function $f \mapsto \int {}^* |f| \, d\mu$ on C''_a is a Riesz seminorm.

Proof. By (21.3),

$$\int {}^* |\rho f| \, d\mu = \int {}^* |\rho| \, |f| \, d\mu = |\rho| \int {}^* |f| \, d\mu.$$

By (21.4) and (21.6),

$$\int {}^* |f+g| \, d\mu \ \leq \ \int {}^* (|f| + |g|) d\mu = \int {}^* |f| \, d\mu + \int {}^* |g| \, d\mu.$$

And by (21.4),

$$|g| \ \leq \ |f| \text{ implies } \int {}^* |g| \, d\mu \ \leq \ \int {}^* |f| \, d\mu.$$

$$\text{QED}$$

Remark. The seminorm $\int {}^* |f| \, d\mu$ is the tool used by Bourbaki [1] and by Stone [6] to develop the Lebesgue integral (they worked in the space of all functions on a set, not just the bounded ones).

We restate (19.4)

(21.12) For a sequence $\{f_n\}$ in $(C_a'')_+$, the following are equivalent:

$1°$ $\lim_n \int {}^*f_n \wedge \mathbf{1}_a d\mu = 0,$

$2°$ $\lim_n \int {}^*(\mathbf{1}_a)_{(f_n - \lambda \mathbf{1}_a)^+} d\mu = 0$ for every $\lambda > 0.$

If one, hence also the other, of these statements holds, then $f_n \longrightarrow 0$ in μ-measure.

And the comment preceding (19.8) becomes: If $\{f_n\}$ is bounded, then the statement "$f_n \longrightarrow f$ in μ-measure" reduces to $\lim_n \int {}^* |f_n - f| \, d\mu = 0.$

We close this § with a classical example to show that the inequalities in this § and in §16 are strict in general.

Let X be the real interval $[0,1]$ and μ the Lebesgue measure. Choose two subsets P, Q of X such that $P \cap Q = \emptyset$, $P \cup Q = X$, and each has outer Lebesgue measure 1. Let d and e be the characteristic elements in C_a'' of P and Q respectively. Thus d and e are components of $\mathbf{1}_a$ satisfying:

$$d \wedge e = 0, \quad d + e = \mathbf{1}_a,$$
$$\int {}^*d \cdot d\mu = \int {}^*e d\mu = 1,$$
$$\int_* d \cdot d\mu = \int_* e d\mu = 0.$$

(I) In (21.6), the last inequality of (1) need not be equality:

$$\int {}^*(d + e) d\mu = \int {}^*\mathbf{1}_a d\mu = \langle \mathbf{1}, \mu \rangle = 1,$$
$$\int {}^*d \cdot d\mu + \int {}^*e d\mu = 2.$$

It follows the inequality $\int^* f d\mu - \int^* g d\mu \leq \int^*(f-g)d\mu$ in (2) there need not be equality. These in turn give us that the inequalities in (16.6) from which those in (21.6) were obtained need not be equalities.

(II) The inequality $|f^*| \leq |f|^*$ in (16.16) needs not be equality: Set $f = d - 2e$, so $d = f^+$, $2e = f^-$. It suffices to show that $\langle |f^*|, \mu \rangle \neq \langle |f|^*, \mu \rangle$. On the one hand, from (16.15), $f^* = d^* - (2e)_* = d^* - 0 > 0$, so $\langle |f^*|, \mu \rangle = \langle f^*, \mu \rangle = 1$. On the other hand, $|f| = d \vee 2e \geq 2e$, so $|f|^* \geq 2e^*$, so $\langle |f|^*, \mu \rangle \geq \langle 2e^*, \mu \rangle = 2$.

(III) The inequalities in (16.17) need not be equalities:
$|(\mathbf{1}_a)^* - e^*| = (\mathbf{1}_a)^* - e^* = \mathbf{1} - e^*$, so $\langle |(\mathbf{1}_a)^* - e^*|, \mu \rangle = 1 - 1 = 0$.
But $|\mathbf{1}_a - e|^* = (\mathbf{1}_a - e)^* = d^*$, so $\langle |\mathbf{1}_a - e|^*, \mu \rangle = 1$.

(IV) The identities obtained by interchanging $\vee$ and $\wedge$ in (16.8), even for two elements, are false: $d \wedge e = 0$, so $(d \wedge e)^* = 0$; we show that $d^* \wedge e^* \neq 0$. $0 \leq d^* \leq \mathbf{1}$, so

(i) $0 \leq (d^*)_\mu \leq \mathbf{1}_\mu$.

And $\langle d^*, \mu \rangle = 1$, so

(ii) $\langle (d^*)_\mu, \mu \rangle = 1$.

It follows from (i) and (ii) that $(d^*)_\mu = \mathbf{1}_\mu$. Similarly, $(e^*)_\mu = \mathbf{1}_\mu$. Hence

$$(d^* \wedge e^*)_\mu = (d^*)_\mu \wedge (e^*)_\mu = \mathbf{1}_\mu \neq 0.$$

Thus $d^* \wedge e^* \neq 0$.

§22. $\int f d\mu$

In this §, μ is still a fixed element of C'_+.

Given $f \in C''_a$, if $\int_* f d\mu = \int^* f d\mu$, we will say that f is μ-integrable *in the standard sense*. We will denote the above common value by $\int f d\mu$, and call it the μ-*integral* of f *in the standard sense*.

Our definition coincides with the standard one for a bounded function on X:

(22.1) For $f \in C''_a$, the following are equivalent:

   1° f is μ-integrable *in the standard sense*,

   2° $\inf\{\langle h, \mu \rangle : h \in U, \ h_a \geq f\} = \sup\{\langle g, \mu \rangle : g \in U, \ g_a \leq f\}$,

   3° $\inf\{\langle h, \mu \rangle : h$ an lsc element, $h_a \geq f\} = \sup\{\langle g, \mu \rangle : g$ a usc

element, $g_a \leq f\}$.

And the common values in 2° and 3° are equal to $\int f d\mu$.

This is an immediate consequence of (21.2).

(22.2) For $f \in C''_a$, the following are equivalent:

   1° f is μ-integrable *in the standard sense*,

   2° $f \in \mathfrak{L}^\infty(\mu)_a$.

This follows from (17.6).

Also from (17.6), we have

(22.3) For $f \in C_a''$, the following are equivalent:

$1°$ $\int^* |f| \, d\mu = 0$,

$2°$ $f \in \mathcal{N}(\mu)_a$.

(22.4) (1) $\int \cdot \, d\mu$ is a positive linear functional on $\mathfrak{L}^\infty(\mu)_a$.

(2) For every $f \in \mathfrak{L}^\infty(\mu)_a$,

$$|\int f d\mu| \leq \int |f| \, d\mu \leq \|f\| \, \|\mu\|.$$

Proof. (1) follows from (21.3), (21.4), and (21.6).

$$(2) \, |\int f d\mu| \leq \int |f| \, d\mu \qquad \text{(from (1))}$$
$$= \langle |f|^*, \mu \rangle$$
$$\leq \| \, |f|^* \, \| \, \|\mu\|$$
$$= \| \, |f| \, \| \, \|\mu\|$$

(for the last equality, cf. the comment following (16.17)).

QED

In $\mathfrak{L}^\infty(\mu)_a$, (21.12) becomes

(22.5) For a sequence $\{f_n\}$ in $\mathfrak{L}^\infty(\mu)_a$, $\{f_n\} \geq 0$, the following are equivalent:

$1°$ $\lim_n \int f_n \wedge \mathbf{1}_a d\mu = 0$

$2°$ $\lim_n \int (\mathbf{1}_a)_{(f_n - \lambda \mathbf{1}_a)} + d\mu = 0$ for every $\lambda > 0$.

If one, hence also the other, of these statements holds, then $f_n \longrightarrow 0$ in μ-measure.

And the statement that a sequence $\{f_n\}$ in $\mathfrak{L}^\infty(\mu)_a$ converges to $f \in \mathfrak{L}^\infty(\mu)$ in μ-measure becomes $\lim_n \int |f_n - f| \wedge \mathbf{1}_a d\mu = 0$. And, finally, the comment preceding (19.8) becomes (for our fixed μ)

(22.6)   For a sequence $\{f_n\}$ in $\mathfrak{L}^\infty(\mu)_a$ bounded μ-a.e., and $f \in \mathfrak{L}^\infty(\mu)_a$, the following are equivalent:

$1°$ $f_n \longrightarrow f$ in μ-measure,

$2°$ $\lim_n \int |f_n - f| d\mu = 0$.

Proof. Assume $1°$. By (19.3), $\{(f_n)^*\}$ is also bounded μ-a.e.; and by (19.6), $(f_n)^* \longrightarrow f^*$ in μ-Measure. Hence by (11.4),

$$\lim_n \langle \, | \, (f_n)^* - f^* \, | \, , \mu \rangle = 0.$$

It follows from (20.9) that this can be written $\lim_n \langle \, | \, f_n - f \, | \, ^*, \mu \rangle = 0$, that is, $\lim_n \int |f_n - f| d\mu = 0$. That $2°$ implies $1°$ follows from (22.5) above.

QED

(22.7) The linear functional $\int \cdot \, d\mu$ is continuous on $\mathfrak{L}_a^\infty$ with respect to the following modes of sequential convergence:

(1) norm convergence,

(2) order convergence,

(3) order convergence μ-a.e.,

(4) convergence in μ-measure of sequences bounded μ-a.e..

Proof. (1) implies (2), (2) implies (3), and, by (20.13), (3) implies (4). Thus we need only prove continuity with respect to (4). But this is the content of (22.6) above.

QED

Note that the continuity with respect to (1), even for nets, follows from (22.4).

From (21.8), we have

(22.8) Given a set $\{f_\alpha\}$ in $\mathfrak{L}^\infty(\mu)_a$, if $g = \wedge_\alpha f_\alpha$ μ-a.e. and $h = \vee_\alpha f_\alpha$ μ-a.e., then

$$\int g d\mu \leq \inf_\alpha \int f_\alpha d\mu \leq \sup_\alpha \int f_\alpha d\mu \leq \int h d\mu.$$

We return to our general band J of C′. The following is clear. $f \in \mathfrak{L}_a^\infty$ if and only if it is ν-integrable for every $\nu \in J_+$. $f \in \mathcal{N}_a$ if and only if $\int f d\nu = 0$ for every $\nu \in J_+$. For $\{f_n\} \subset \mathfrak{L}_a^\infty$, $f_n \longrightarrow f$ in measure if and only if the conditions in (22.5) hold for every $\nu \in J_+$. The functions $\{\int \cdot d\nu \colon \nu \in J_+\}$ are positive linear functionals on $\mathfrak{L}_a^\infty$ continuous with respect to convergence of sequences in each of the modes in (22.7). Finally, $\mathfrak{L}_a^\infty$ and J are in duality under the bilinear form

$$\langle\langle f, \nu \rangle\rangle = \int f d\nu \quad (f \in \mathfrak{L}_a^\infty, \ \nu \in J_+)$$

extended to all of J by linearity.

§23. The classical theorems

In this final §, we obtain from the theorems in Chapter 3, the corresponding ones in the standard Lebesgue theory on X.

(13.1) gives us the equivalence for bounded functions of integrability and measurability:

(23.1) For $f \in C_a''$, the following are equivalent:

$1°$ $f \in \mathfrak{L}_a^\infty$,

$2°$ $(\mathbf{1}_a)_{(f-\lambda \mathbf{1}_a)^+} \in \mathfrak{L}_a^\infty$ for every $\lambda \in \mathbb{R}$.

Proof. By (17.3), $1°$ is equivalent to

(i) $f^* \in \mathfrak{L}^\infty$.

And by the Lemma in the proof of (19.4) together with (17.3), $2°$ is equivalent to

(ii) $\mathbf{1}_{(f^*-\lambda 1)^+} \in \mathfrak{L}^\infty$ for every $\lambda \in \mathbb{R}$.

By (13.1), (i) and (ii) are equivalent.

QED

(21.9) gives us

(23.2) (Fatou's Lemma) Given $\mu \in C'_+$, a sequence $\{f_n\}$ in $\mathfrak{L}^\infty(\mu)_a$, $g = \liminf_n (f_n)$ μ-a.e., and $h = \limsup_n (f_n)$ μ-a.e.. Then

$$\int g d\mu \leq \liminf_n \int f_n d\mu \leq \limsup_n \int f_n d\mu \leq \int h d\mu.$$

Remark.   It   follows   that   if $f_n \to f$ μ-a.e.,   then $\int f d\mu = \lim_n \int f_n d\mu$. This also follows from (22.6) and (22.7).

(23.3) (Riesz) Given $\mu \in C'_+$ and a sequence $\{f_n\}$ in $\mathfrak{L}^\infty(\mu)_a$. If $f_n \to f$ in μ-measure, then some subsequence order converges unboundedly μ-a.e. to f.

Proof. By (19.6), $(f_n)^* \to f^*$ in μ-Measure. Hence, by (13.3), there exists a subsequence $\{f_{n_m}\}$ such that $(f_{n_m})^* \xrightarrow{u} f^*$ μ-a.e. (as $m \to \infty$). It follows from (20.12) that $f_{n_m} \xrightarrow{u} f$ μ-a.e..

QED

The Lebesgue Bounded Convergence Theorem is a corollary of the following result, which we have already established.

(23.4) Let $\{f_n\}$ be a sequence of $\mathfrak{L}^\infty_a$ bounded a.e.. If $f_n \to f$ in measure, then $f \in \mathfrak{L}^\infty_a$ and

$$\int f d\nu = \lim_n \int f_n d\nu \text{ for every } \nu \in J_+.$$

This is contained in (20.1) and (22.7) (cf. the final paragraph in §22).

(23.5) (First Egorov Theorem) Given $\nu \in C'_+$ and $\{f_n\} \subset \mathfrak{L}^\infty(\mu)_a$. If $f_n \xrightarrow{u} f$ μ-a.e., then for each $\delta > 0$, there exist components d, e of $\mathbf{1}_a$ such that:

(1) $d, e \in \mathfrak{L}^\infty(\mu)_a$,

(2) $d + e = \mathbf{1}_a$,

(3) $\int d \cdot d\mu \leq \delta$,

(4) $\lim_n \| (f_n)_e - f_e \| = 0$.

Moreover, if the f_n's and f all lie in a band G of C''_a such that $\mathbf{1}_G \in \mathfrak{L}^\infty(\mu)_a$, then the conclusion holds with $\mathbf{1}_a$ replaced by $\mathbf{1}_G$.

Proof. By (20.12), $(f_n)^* \xrightarrow{u} f^*$ μ-a.e.. Fix δ and let d_0, e_0 be given by (14.2) for this convergence. (It is clear from the proof there that, since $\{(f_n)^*\} \subset \mathfrak{L}^\infty(\mu)$, d_0 and e_0 lie in $\mathfrak{L}^\infty(\mu)$.) $(d_0)_a$ and $(e_0)_a$ are easily seen to have the desired properties.

QED

(23.6) (Second Egorov Theorem) Given $\mu \in C'_+$ and $\{f_n\} \subset \mathfrak{L}^\infty(\mu)_a$. If $f_n \xrightarrow{u} f$ μ-a.e., then there exist components of $\mathbf{1}_a$, $\{e_m: m=0,1,2,...\}$ in $\mathfrak{L}^\infty(\mu)_a$ such that:

(1) $e_{m_1} \wedge e_{m_2} = 0$ for $m_1 \neq m_2$,

(2) $\bigvee_m e_m = \mathbf{1}_a$,

(3) $\int e_o d\mu = 0$,

(4) $\lim_n \| (f_n)_{e_m} - f_{e_m} \| = 0$ for m=1, 2, ...

And if the f_n's and f all lie in a band G of C''_a such that $\mathbf{1}_G \in \mathfrak{L}^\infty(\mu)_a$, then the conclusion holds with $\mathbf{1}_a$ replaced by $\mathbf{1}_G$.

This follows from (23.5) in the same way that (14.3) followed from (14.2)

(23.7) (Lusin) Given $\mu \in C'_+$, then for $f \in C''_a$, the following are equivalent:

1° $f \in \mathfrak{L}^\infty(\mu)_a$;

2° there exists a bounded sequence $\{f_n\}$ in C_a such that $f_n \longrightarrow f$ μ-a.e.;

3° for each $\delta > 0$, there is a component e of $\mathbf{1}_a$, uppersemicontinuous on X (so a fortiori in $\mathfrak{L}^\infty(\mu)_a$) such that

(a) $\int (\mathbf{1}_a - e) d\mu \leq \delta$,

(b) $f_e \in C_e$;

4° there exists a sequence $\{e_n\}$ of components of $\mathbf{1}_a$, each uppersemicontinuous on X, such that

(a) $e_n \uparrow \mathbf{1}_a$ μ-a.e.,

(b) $f_{e_n} \in C_{e_n}$ for every n;

5° there exists a sequence $\{e_n\}$ of components of $\mathbf{1}$, all lying in $\mathfrak{L}^\infty(\mu)_a$, such that

(a) $e_n \uparrow \mathbf{1}_a$ μ-a.e.,

(b) $f_{e_n} \in C_{e_n}$ for every n.

Proof. Assume 1°. So there exists $g \in \mathfrak{L}^\infty(\mu)$ such that $g_a = f$. By (15.1), there exists a bounded sequence $\langle g_n \rangle$ in C such that $g_n \longrightarrow g$ μ-a.e.. Set $f_n = (g_n)_a$ (n=1, 2, ...). Since, for every n, $g_n \in C \subset U$, $g_n = (f_n)^*$. Hence, by (20.11), $\{f_n\}$ satisfies 2°.

Assume 2°. $\{f_n\} \subset \mathfrak{L}^\infty(\mu)_a$, so by (20.11), $(f_n)^* \longrightarrow f^*$ μ-a.e.. It follows from (15.1) that given $\delta > 0$, there exists a usc component e_0 of $\mathbf{1}$ such that:

(i) $\langle \mathbf{1} - e_0, \mu \rangle \leq \delta$,

(ii) $(f^*)_{e_0} \in C_{e_0}$.

Then $e = (e_0)_a$ satisfies 3°.

We leave the remainder of the proof to the reader.

QED

References

1. Bourbaki, N., INTEGRATION, Chapters I–IV, Hermann et Cie., 1952.

2. DeMarr, R., *Partially ordered linear spaces and locally convex linear topological spaces*, Ill. Jour. Math. 8 (1964), 601–606.

3. Kaplan, S., THE BIDUAL OF C(X) I, North Holland, 1985.

4. _____, *On unbounded order convergence*, Submitted.

5. Nakano, H., *Ergodic theorems in semi-ordered linear spaces*, Ann. Math. 49 (1948), 538–556.

6. Stone, M.H., *Notes on integration*, Proc. Nat. Acad. Sci. 34 (1948), 336–342, 447–455, 483–490 and 35 (1949), 50–58.

Index of Terminology

Index of Symbols

Symbols not listed below are defined in our monograph [3].

Department of Mathematics
University of North Carolina
Chapel Hill, N.C. 27599 – 3250
U.S.A.

Editorial Information

To be published in the *Memoirs*, a paper must be correct, new, nontrivial, and significant. Further, it must be well written and of interest to a substantial number of mathematicians. Piecemeal results, such as an inconclusive step toward an unproved major theorem or a minor variation on a known result, are in general not acceptable for publication. *Transactions* Editors shall solicit and encourage publication of worthy papers. Papers appearing in *Memoirs* are generally longer than those appearing in *Transactions* with which it shares an editorial committee.

As of January 31, 1996, the backlog for this journal was approximately 5 volumes. This estimate is the result of dividing the number of manuscripts for this journal in the Providence office that have not yet gone to the printer on the above date by the average number of monographs per volume over the previous twelve months, reduced by the number of issues published in four months (the time necessary for preparing an issue for the printer). (There are 6 volumes per year, each containing at least 4 numbers.)

A Copyright Transfer Agreement is required before a paper will be published in this journal. By submitting a paper to this journal, authors certify that the manuscript has not been submitted to nor is it under consideration for publication by another journal, conference proceedings, or similar publication.

Information for Authors and Editors

Memoirs are printed by photo-offset from camera copy fully prepared by the author. This means that the finished book will look exactly like the copy submitted.

The paper must contain a *descriptive title* and an *abstract* that summarizes the article in language suitable for workers in the general field (algebra, analysis, etc.). The *descriptive title* should be short, but informative; useless or vague phrases such as "some remarks about" or "concerning" should be avoided. The *abstract* should be at least one complete sentence, and at most 300 words. Included with the footnotes to the paper, there should be the 1991 *Mathematics Subject Classification* representing the primary and secondary subjects of the article. This may be followed by a list of *key words and phrases* describing the subject matter of the article and taken from it. A list of the numbers may be found in the annual index of *Mathematical Reviews*, published with the December issue starting in 1990, as well as from the electronic service e-MATH [**telnet e-MATH.ams.org** (or **telnet 130.44.1.100**). Login and password are **e-math**]. For journal abbreviations used in bibliographies, see the list of serials in the latest *Mathematical Reviews* annual index. When the manuscript is submitted, authors should supply the editor with electronic addresses if available. These will be printed after the postal address at the end of each article.

Electronically prepared papers. The AMS encourages submission of electronically prepared papers in $\mathcal{A}\mathcal{M}\mathcal{S}$-TeX or $\mathcal{A}\mathcal{M}\mathcal{S}$-LaTeX. The Society has prepared author packages for each AMS publication. Author packages include instructions for preparing electronic papers, the *AMS Author Handbook*, samples, and a style file that generates the particular design specifications of that publication series for both $\mathcal{A}\mathcal{M}\mathcal{S}$-TeX and $\mathcal{A}\mathcal{M}\mathcal{S}$-LaTeX.

Authors with FTP access may retrieve an author package from the Society's Internet node **e-MATH.ams.org** (130.44.1.100). For those without FTP access, the author package can be obtained free of charge by sending e-mail to `pub@math.ams.org` (Internet) or from the Publication Division, American Mathematical Society, P.O. Box 6248, Providence, RI 02940-6248. When requesting an author package, please specify $\mathcal{AMS}$-TEX or $\mathcal{AMS}$-LATEX, Macintosh or IBM (3.5) format, and the publication in which your paper will appear. Please be sure to include your complete mailing address.

Submission of electronic files. At the time of submission, the source file(s) should be sent to the Providence office (this includes any TEX source file, any graphics files, and the DVI or PostScript file).

Before sending the source file, be sure you have proofread your paper carefully. The files you send must be the EXACT files used to generate the proof copy that was accepted for publication. For all publications, authors are required to send a printed copy of their paper, which exactly matches the copy approved for publication, along with any graphics that will appear in the paper.

TEX files may be submitted by email, FTP, or on diskette. The DVI file(s) and PostScript files should be submitted only by FTP or on diskette unless they are encoded properly to submit through e-mail. (DVI files are binary and PostScript files tend to be very large.)

Files sent by electronic mail should be addressed to the Internet address `pub-submit@math.ams.org`. The subject line of the message should include the publication code to identify it as a Memoir. TEX source files, DVI files, and PostScript files can be transferred over the Internet by FTP to the Internet node `e-math.ams.org` (130.44.1.100).

Electronic graphics. Figures may be submitted to the AMS in an electronic format. The AMS recommends that graphics created electronically be saved in Encapsulated PostScript (EPS) format. This includes graphics originated via a graphics application as well as scanned photographs or other computer-generated images.

If the graphics package used does not support EPS output, the graphics file should be saved in one of the standard graphics formats—such as TIFF, PICT, GIF, etc.—rather than in an application-dependent format. Graphics files submitted in an application-dependent format are not likely to be used. No matter what method was used to produce the graphic, it is necessary to provide a paper copy to the AMS.

Authors using graphics packages for the creation of electronic art should also avoid the use of any lines thinner than 0.5 points in width. Many graphics packages allow the user to specify a "hairline" for a very thin line. Hairlines often look acceptable when proofed on a typical laser printer. However, when produced on a high-resolution laser imagesetter, hairlines become nearly invisible and will be lost entirely in the final printing process.

Screens should be set to values between 15% and 85%. Screens which fall outside of this range are too light or too dark to print correctly.

Any inquiries concerning a paper that has been accepted for publication should be sent directly to the Editorial Department, American Mathematical Society, P. O. Box 6248, Providence, RI 02940-6248.

Other Titles in This Series

(Continued from the front of this publication)

(See the AMS catalog for earlier titles)